Lecture Notes in Mathematics

Edited by A. Dold and B. Eckmann

504

Serge Lang
Hale Trotter

Frobenius Distributions
in GL_2-Extensions

Distribution of Frobenius Automorphisms in
GL_2-Extensions of the Rational Numbers

Springer-Verlag
Berlin · Heidelberg · New York 1976

Authors
Serge Lang
Mathematics Department
Yale University
New Haven, Connecticut 06520
USA

Hale Freeman Trotter
Fine Hall
Princeton University
Princeton, New Jersey 08540
USA

Library of Congress Cataloging in Publication Data

Lang, Serge, 1927-
   Frobenius distributions in GL_2-extensions.

   (Lecture notes in mathematics ; 504)
   Bibliography: p.
   Includes index.
   1. Probabilistic number theory. 2. Galois theory.
3. Field extensions (Mathematics) 4. Numbers,
Rational. I. Trotter, Hale F., joint author. II. Ti-
tle. III. Series: Lecture notes in mathematics
(Berlin) ; 504.
QA3.L28 no. 504   [QA241.7]   510'.8s [512'.7]
                                          75-45242

AMS Subject Classifications (1970): 10 K 99, 12 A 55, 12 A 75, 33 A 25

ISBN 3-540-07550-X Springer-Verlag Berlin · Heidelberg · New York
ISBN 0-387-07550-X Springer-Verlag New York · Heidelberg · Berlin

This work is subject to copyright. All rights are reserved, whether the whole or part of the material is concerned, specifically those of translation, reprinting, re-use of illustrations, broadcasting, reproduction by photocopying machine or similar means, and storage in data banks.
Under § 54 of the German Copyright Law where copies are made for other than private use, a fee is payable to the publisher, the amount of the fee to be determined by agreement with the publisher.
© by Springer-Verlag Berlin · Heidelberg 1976
Printed in Germany
Offsetdruck: Julius Beltz, Hemsbach/Bergstr.

ACKNOWLEDGMENTS

Both Lang and Trotter are supported by NSF grants. The final draft of this monograph was written while Lang was at the Institute for Advanced Study, whose hospitality we appreciated.

We also thank Mrs. Helen Morris for a superb job of varityping the final copy.

INTRODUCTION

We are interested in a distribution problem for primes related to elliptic curves, but which can also be described solely in terms of the distribution of Frobenius elements in certain Galois extensions of the rationals. We therefore first describe this situation, and then indicate its connection with elliptic curves.

Let K be a Galois (infinite) extension of the rationals, with Galois group G. We suppose given a representation

$$\rho : G \longrightarrow \prod GL_2(Z_\ell)$$

which we assume gives an embedding of G onto an open subgroup of the product, taken over all primes ℓ. We let

$$\rho_\ell : G \longrightarrow GL_2(Z_\ell)$$

be the projection of ρ on the ℓ-th factor. We assume that there is an integer Δ such that if p is a prime and $p \nmid \Delta\ell$, then p is unramified in ρ_ℓ, or in other words, the inertia group at a prime of K above p is contained in the kernel of ρ_ℓ. Then the Frobenius class σ_p is well defined in the factor group

$$G_\ell = G/\mathrm{Ker}\, \rho_\ell \,,$$

and $\rho_\ell(\sigma_p)$ has a characteristic polynomial which we assume of the form

$$X^2 - t_p X + p \,.$$

We assume that t_p is an integer independent of ℓ, and call t_p the trace of Frobenius. Finally we assume that the roots of the characteristic polynomial have absolute value $\sqrt{p}$, and are complex conjugates of each other. Let π_p be such a root.

Let t_0 be a given integer. Let k be a given imaginary quadratic field. We let $N_{t_0,\rho}(x)$ be the number of primes $p \leq x$ such that $t_p = t_0$. We let $N_{k,\rho}(x)$

be the number of primes $p \leq x$ such that $Q(\pi_p) = k$. If $k = k_D$ is the field

$$k_D = Q(\sqrt{D})$$

with discriminant D, we also write $N_D(x)$, $N_{D,\rho}(x)$ or $N_k(x)$ instead of $N_{k,\rho}(x)$.

We conjecture that there are constants $C(t_0, \rho)$ and $C(k, \rho) > 0$ such that we have the asymptotic relations

$$N_k(x) \sim C(k,\rho) \frac{\sqrt{x}}{\log x} \quad \text{and} \quad N_{t_0}(x) \sim C(t_0, \rho) \frac{\sqrt{x}}{\log x}.$$

The constants depend on k, ρ or t_0, ρ respectively. If $C(t_0, \rho) = 0$ then the asymptotic relation is to be interpreted to mean that $N_{t_0}(x)$ is bounded. If $C(t_0, \rho) \neq 0$, then the asymptotic relation has the usual meaning. We shall also see that $C(0, \rho) \neq 0$. Cf. Part I, §4, Remark 1.

Actually, define

$$\pi_{\frac{1}{2}}(x) = \sum_{p \leq x} \frac{1}{2\sqrt{p}}.$$

This is essentially

$$\int^x \frac{1}{2\sqrt{x}} d\pi(x) \sim \int^x \frac{1}{2\sqrt{x}} \frac{dx}{\log x},$$

which can be integrated by parts with $u = 1/\log x$ and $dv = \frac{dx}{2\sqrt{x}}$ to show that

$$\pi_{\frac{1}{2}}(x) \sim \frac{\sqrt{x}}{\log x}.$$

Both in the theoretical analysis and in the numerical computations, it is the asymptotic relations

$$N_k(x) \sim C(k,\rho) \pi_{\frac{1}{2}}(x) \quad \text{and} \quad N_{t_0}(x) \sim C(t_0,\rho) \pi_{\frac{1}{2}}(x)$$

which arise. Therefore it is more natural to deal with $\pi_{\frac{1}{2}}(x)$, rather than with the elementary form

$$\frac{\sqrt{x}}{\log x},$$

which converges asymptotically only slowly to $\pi_{\frac{1}{2}}(x)$.

Our arguments to make the conjecture plausible involve only the Galois representation ρ, the Tchebotarev and Hecke distribution theorems in finite Galois extensions, and a conjectured distribution function for the angles of Frobenius elements. One may view our study as a first attempt to formulate for certain infinite extensions distribution laws for Frobenius elements. On the other hand, the motivation also arises from the theory of elliptic curves as follows.

Let A be an elliptic curve over the rationals. Let $K = Q(A_{tor})$ be the field obtained by adjoining the coordinates of its torsion points. Then the Galois group admits a natural representation in $\prod GL_2(Z_\ell)$. We assume that A has no complex multiplication. We then know from Serre's work [S 2] that the representation is open in the product. It is also known that the other properties mentioned above are satisfied (especially that the roots of the characteristic polynomial are complex conjugates of each other, which is none other than the Riemann Hypothesis, Hasse's Theorem). When the representation arises from an elliptic curve, we then write also $N_{k,A}(x)$, etc., replacing ρ by A in the notation. We note that the constants $C(k,A)$ and $C(t_0,A)$ are obviously isogeny invariants of A. (Isogenies over the algebraic closure of Q are allowed.)

A prime p is called supersingular when $t_p = 0$. This is a standard interesting case in the theory of elliptic curves. There are numerous other characterizations of this case, which are however irrelevant for us here (cf. for instance Appendix 2 of [L 1]). Serre had already observed that the densities of supersingular primes, or those for which $Q(\pi_p) = k$, are zero ([S 1] for the supersingular case, private communication in the other). Mazur emphasized the importance of the case when $t_p = 1$ for the arithmetic of elliptic curves (see [Ma], Propositions 8.5 and 8.14) and called the prime p anomalous when $t_p = 1$. In the case of complex multiplication, if a prime is anomalous, then it lies in a quadratic progression, and the conjectured distribution of such primes can be reduced to a conjecture of Hardy-Littlewood, that it is of the form

$$C \frac{\sqrt{x}}{\log x}$$

for some constant C. The Galois group of a curve with complex multiplication is of course not a GL_2-group, and our situation is more complicated.

(Strictly speaking, one should define supersingular (resp. anomalous) by the condition $t_p \equiv 0$ (resp. $t_p \equiv 1$) mod p, but since $|t_p| < 2\sqrt{p}$, this amounts to the same thing for primes > 5, so the distinction is irrelevant for our purpose, which is to count primes asymptotically.)

The axiomatization of the distribution properties only in terms of the Galois group is important for eventual applications to representations arising from modular forms other than those associated with elliptic curves. One knows from the work of Swinnerton-Dyer [SwD] that they give rise to (essentially) GL_2-extensions of the rational numbers. (Cf. also Ribet [R].) The characteristic polynomial of a Frobenius element is of the form

$$X^2 - t_p X + p^{k-1} = 0.$$

The analogue of our $\sqrt{p}$ is then $p^{(k-1)/2}$. This leads us to think that when $k \geq 4$, there is only a finite number of primes such that the Frobenius element belongs to the given quadratic field, or such that $t_p = 0$. This would be in line with the Lehmer conjecture that $\tau_p \neq 0$ for all p, where τ_p is the trace of Frobenius for the best known cusp form Δ from the theory of elliptic functions. For $k = 3$, one gets an intermediate asymptotic behavior, and for $k = 1$ one gets back to the oldest situation of actual densities, since the associated Galois group is finite. The case with $k > 2$ introduces enough perturbations in our arguments that we shall handle it elsewhere.

From a naive approach, one already suspects that the asymptotic behavior has something to do with $\pi_{\frac{1}{2}}(x)$. Indeed, the trace of Frobenius t_p must lie in the interval $|t_p| < 2\sqrt{p}$. Under equal probability that it hits any integer in this interval, this probability is $\frac{1}{4\sqrt{p}}$. Summing for $p \leq x$ yields the $\pi_{\frac{1}{2}}(x)$. In reality, the probabilities of hitting the different integers in the interval are not equal, but depend in a fairly complicated way on the Galois representation. In the imaginary quadratic case one wants the probability of coincidence of t_p with the trace of some integer of the field k with norm p. This probability involves an interaction between the field of division points and the maximal abelian extension of k, and becomes especially complicated when the intersection of these two fields is larger (as it may be, by a finite extension) than the field of all roots of unity over the rationals. The effect that this last complication can have on the probabilistic factor is one of the more interesting things we have encountered in the present study.

Tuskina [Tu] already conjectured an equivalent asymptotic formula for the distribution of supersingular primes, purely on the basis of empirical evidence (but without making any conjecture as to the value of the constant).

The computation of the constant makes it necessary to have an exact description of the Galois groups. This can be an arduous task. We obviously rely heavily on Serre [S 2], and also use ideas of Shimura [Sh], especially in determining the group of $X_0(11)$.

Our heuristic method is to consider probabilistic models in which we consider the sequence of traces of Frobenius $\{t_p\}$ to be a random sequence. We choose the simplest model for which almost all sequences have asymptotic properties consistent with the density theorems of Tchebotarev and Hecke, concerning the distribution of primes with given element of the Galois group, and in sectors of the plane, and also consistent with the Sato-Tate conjecture. We show that for this model, almost all sequences have an asymptotic behavior of the form mentioned (a constant times $\pi_{\frac{1}{2}}(x)$), and we compute this constant explicitly in terms of the Galois group. Our conjecture is that the sequence of Frobenius elements has this behavior. More precisely, say in the supersingular case, this amounts to saying that the probability that p is supersingular is asymptotic to $C(0,\rho) \cdot \frac{1}{2\sqrt{p}}$, and similarly in the other cases, using $C(k,\rho)$ instead of $C(0,\rho)$.

In the case of the quadratic field k with discriminant D, the constant is inversely proportional to $\sqrt{|D|}$, and can be expressed as a product of local factors depending on ℓ and D for almost all primes ℓ, as well as a factor depending on the special position of the Galois group in the product at a finite number of exceptional primes depending on ρ, and D. There is also a factor at infinity, derived from the Sato-Tate distribution.

The factors at finite primes can be expressed as integrals over ℓ-adic sets of certain functions which are Harish transforms. We develop ab ovo the theory of Harish transforms, which can be formulated completely naively in terms of the direct image of Haar measure under the trace-determinant map (i.e. the map which associates with each matrix the coefficients of its characteristic polynomial). The theory of this transform has independent interest, and is given in Part II, §7 and §8.

Our axiomatization involving only a GL_2-Galois extension of the rationals gives rise to various questions.

1. Are there such extensions (all of elliptic type, see §1 below) other than those arising from the division points of an elliptic curve?

This question does not seem to fit exactly in the general Langlands framework, since no assumption is made here about the associated zeta function of the GL_2-extension.

2. Cusp forms do not appear in the present work. Can the conjecture be even remotely approached for the case of elliptic curves by using explicit formulas for the coefficients of the associated cusp form, e.g. formulas as in Manin [Man]? How do the congruence conditions and the finite part of the constant arising from the Galois representation translate into conditions on the coefficients of the associated cusp form? How can one describe only in terms of these coefficients the conditions which determine the "fixed trace progression," or the "given imaginary quadratic field progression"?

In some sense what we are about is to reconstruct the arithmetic of an elliptic curve without complex multiplication by piecing together the totality of elliptic curves with complex multiplication in a certain way. There should be something like a reciprocity law which bears to our conjectured asymptotic behavior a relationship analogous to that which the Artin reciprocity law bore to the Frobenius conjectured density properties, proved by Tchebotarev.

3. Again in the case of elliptic curves, can one give a condition on the analytic behavior of the associated Dirichlet series (zeta function) which implies our conjectured asymptotic property? In particular, is there significance to the partial Euler products taken over those p which are supersingular, or which correspond to a given imaginary quadratic field, and is there an L-series formalism attached to such products? The Hardy-Littlewood paper [H−L] is in two parts. The first shows how various Riemann Hypotheses imply distribution results. The second, including the conjecture on primes in quadratic progressions, limits itself to heuristic arguments. Therefore, even in that case, it would be interesting to see what analytic properties of zeta functions imply the conjectured asymptotic behavior.

4. Adapting to the present situation the classical view point of characterizing Galois extensions by those primes that split completely, it is reasonable to expect that two elliptic curves over the rationals are isogenous if they have the same set of supersingular primes, except possibly for a subset having an asymptotic order of magnitude strictly smaller than $\sqrt{x}/\log x$. Further comments are made on this in §4, when we have more precise definitions to discuss the matter technically.

For simplicity of expression, the conjecture may be weakened by requiring that the two curves have the same sets of supersingular primes, except for a finite number. In §4 we shall see that it may be strengthened by supposing merely that the common set of supersingular primes not be $O(\log \log x)$.

The elliptic curve A has a rational invariant j_A. For all non-exceptional primes p, we have

$Q(\pi_p) = k$ *if and only if* $j_A \equiv j(\mathfrak{o}) \pmod{\mathfrak{p}}$ *for some order* $\mathfrak{o}$ *in k and some prime* $\mathfrak{p}$ *over* p *in* $k(j(\mathfrak{o}))$.

According to a theorem of Deuring (cf. [L 1], Chapter 13, §4, Theorem 13), we can pick $\mathfrak{o}$ such that p splits completely in $k(j(\mathfrak{o}))$, and the above congruence condition has to be satisfied. It is standard (cf. [L 1], Chapter 8, §1, Corollary of Theorem 7) that there is only a finite number of imaginary quadratic orders $\mathfrak{o}$ such that $j(\mathfrak{o})$ lies in a given number field. However this approach through an increasing tower of orders and congruence conditions did not seem to lead towards a determination of the asymptotic behavior of the distribution of Frobenius elements belonging to the given quadratic field k.

Finally we observe that in the light of Yoshida's proof of the analogue of the Sato conjecture in the function field case [Y], it is possible that enough is known about the distribution of Frobenius elements in that case to be able to give a proof of the analogue of our conjecture. Of course, there is no question of having infinitely many supersingular values of j, which are necessarily finite, and in characteristic p one can only have imaginary quadratic fields as algebras of endomorphisms in which p splits completely, by Deuring's theorems. Cf. [L 1], Chapter 13 and 14. Except for these limitations, one expects a similar theory to hold. The approach through the congruence values $j(\mathfrak{o})$ as $\mathfrak{o}$ ranges over orders of k with conductor prime to p may in fact work in this case, in line with Ihara's ideas [I] which were used by Yoshida [Y].

There remains to say a few words about the logistics of this paper. In Part I, we discuss the fixed trace case In Part II, we treat the imaginary quadratic distribution. While the finite part of the constant stabilizes at finite level in Part I, it does not in Part II, and its theoretical analysis, as well as practical computation requires a more elaborate discussion. Finally, in Part III, we put together special computations dealing with the quadratic fields for which the GL_2-extension has an intersection with the maximal abelian extension of k which is strictly bigger than the field of all roots of unity. These cases are the most interesting.

For instance, the exceptionally large number of occurrences of $Q(\sqrt{-11})$ for $X_0(11)$ must be reflected in a correspondingly large prediction. (It occurs 88 times, when most other fields occur at most one-fourth this many times.) This requires a description of the field of 4-division points. Other cases require a similar description of the 3-division points. Since we felt that our computations should be checkable by anyone else interested to do so, we have included the full details in all cases.

In Part IV we present and discuss the numerical results for five curves and the first 5,000 primes. For one of the curves, $X_0(11)$, the calculation was pushed to include almost 190,000 primes. On the whole, the fit between actual and predicted values is good. We feel that the data are compatible with the conjecture. There are discrepancies, but they seem to lie within the range of reasonable statistical fluctuations.

PART I

SUPERSINGULAR AND FIXED TRACE DISTRIBUTION

PRELIMINARIES

1. The Galois representation in GL_2	17
2. Some notions of probability	20

THE DISTRIBUTION FOR FIXED TRACE

3. The probabilistic model	29
4. The asymptotic behavior	33

EXAMPLES

5. Serre curves, $M = 2q$, the general formula	41
6. Computations of Galois groups	46
7. The curve $y^2 = x^3 + 6x - 2$	51
8. The Shimura curve $X_0(11)$	55

PART II

IMAGINARY QUADRATIC DISTRIBUTION

Introduction	69

THE FIXED TRACE CASE

1.	Fixed traces from the quadratic field	77
2.	Computation of the constant for fixed trace	84

THE MODEL FOR THE MIXED CASE

3.	The mixed Galois representations	91
4.	The probabilistic model	104
5.	The asymptotic behavior	108
6.	The finite part of the constant as a quotient of integrals	112

COMPUTATIONS OF HARISH TRANSFORMS

7.	Haar measure under the trace-determinant map on Mat_2. General formalism.	123
8.	Relations with the trace-norm map on k	133
9.	Computation of C_ℓ for almost all ℓ	141
10.	The constant for Serre curves, $K \cap k_{ab} = Q_{ab}$	143
11.	The constant for $X_0(11)$	149

PART III

SPECIAL COMPUTATIONS

Introduction — 157

GENERAL LEMMAS

1. Lemmas on commutator subgroups — 163
2. $G_2 = GL_2(\mathbb{Z}_2)$ — 165
3. Cases when $K \cap k_{ab} = \mathbb{Q}_{ab}$ — 174
4. $K \cap k_{ab}$ when $k = \mathbb{Q}(\sqrt{-3})$ and $GL_2(\mathbb{Z}_3)$ splits — 181
5. $K \cap k_{ab}$ in other cases — 185

$k = \mathbb{Q}(\sqrt{-3})$

6. The action of $\mathcal{A}$ on $k(\Delta^{1/3})$ — 191
7. The constant for Serre fiberings, $k = \mathbb{Q}(\sqrt{-3})$, $M = 2q$, q odd prime $\neq 3$, $\Delta = \pm q^n$ — 195
8. Computation of integrals — 201

$k = \mathbb{Q}(i)$

9. The constant for Serre fiberings, q odd $\neq 3$ — 209

$k = \mathbb{Q}(\sqrt{\Delta})$

10. The action of $\mathcal{A}$ on $k(A_2, \Delta^{1/4})$ when $k = \mathbb{Q}(\sqrt{\Delta})$ — 215
11. The action of matrices on $k(A_4)$ — 218
12. Computation of integrals and the constant — 221

PART IV

NUMERICAL RESULTS

SUPERSINGULAR AND FIXED TRACE DISTRIBUTION

1. General discussion of results — 235
2. Tables
   Table I : Fixed trace distributions — 239
   Table II : Supersingular primes — 240
   Table III: Primes with $t_p = 1$ — 241
   Table IV: Traces of Frobenius — 242

IMAGINARY QUADRATIC DISTRIBUTION

3. General discussion of results — 249
4. Tables
   Table V : Imaginary quadratic distributions — 253
   Table VI : Primes associated with fields of small discriminant, for curves A and B — 258
   Table VII: Distribution of primes associated with small discriminants — 260

EXTENDED RESULTS FOR $X_0(11)$

5. Discussion and description of tables — 265
   Table VIII: Supersingular primes — 267
   Table IX : Imaginary quadratic distribution — 268
   Table X : Distribution of primes for fields with small discriminants — 269

Remarks on the Computations — 271

Bibliography — 273

PART I

SUPERSINGULAR AND FIXED TRACE DISTRIBUTION

PRELIMINARIES

§1. The Galois representation in GL_2

We fix a Galois (infinite) extension K of the rationals, with Galois group G, and a representation

$$\rho : G \longrightarrow \prod GL_2(Z_\ell)$$

which we assume gives an embedding of G onto an open subgroup of the product, taken over all primes ℓ. We let

$$\rho_\ell : G \longrightarrow GL_2(Z_\ell)$$

be the projection of ρ on the ℓ-th factor, and let G_ℓ be the factor group

$$G/\mathrm{Ker}\, \rho_\ell .$$

We let K_ℓ be the fixed field of $\mathrm{Ker}\, \rho_\ell$, so that G_ℓ is the Galois group of K_ℓ over Q. A Galois extension of the rationals with a representation of its Galois group as above will be called a GL_2-extension.

For each positive integer M, we may reduce $\prod GL_2(Z_\ell)$ mod M, and thereby obtain a factor representation

$$\rho_{(M)} : G \longrightarrow GL_2(Z/MZ) .$$

The factor group of G by $\mathrm{Ker}\, \rho_{(M)}$ is denoted by $G(M)$, and the fixed field of $\mathrm{Ker}\, \rho_{(M)}$ by $K(M)$. Then $K(M)$ is a finite Galois extension with group $G(M)$. We also call $G(M)$ the reduction of G mod M.

We denote Z/MZ by $Z(M)$, and use similar notation for other objects reduced mod M. For instance, $Z(M)^* = (Z/MZ)^*$ is the group of units in $Z(M)$.

We denote by G_M the projection of $\rho(G)$ into the finite product

$$\prod_{\ell \mid M} GL_2(Z_\ell) ,$$

and use ρ_M to denote the representation of G in this projection. We shall say that M splits ρ if we have an isomorphism

$$\rho(G) = \prod_{\ell \nmid M} GL_2(\mathbb{Z}_\ell) \times G_M .$$

Note that G_M is open in the finite product.

On the other hand, let M be arbitrary, and let

$$r_M : \prod_{\ell \mid M} GL_2(\mathbb{Z}_\ell) \longrightarrow GL_2(\mathbb{Z}(M))$$

be the reduction map. We shall say that M is **stable** if the following condition is satisfied.

ST 1. $\qquad\qquad G_M = r_M^{-1}(G(M)) .$

It then follows at once that the next condition is also satisfied.

ST 2. *For every element $\bar\sigma \in G(M)$, we have*

$$r_M^{-1}(\bar\sigma) = \prod_{\ell \mid M} r_{\ell^{m(\ell)}}^{-1}(\bar\sigma(\ell^{m(\ell)}) ,$$

where $M = \prod \ell^{m(\ell)}$ *is the prime power decomposition of* M, *and* $\bar\sigma(\ell^{m(\ell)})$ *is the reduction of $\bar\sigma$ mod $\ell^{m(\ell)}$.*

We use the notation

$$G(M^\infty) = \lim_{n \to \infty} G(M^n) .$$

Then $G_M = G(M^\infty)$, and $G(M^\infty)$ is completely determined by $G(M)$ if M is stable. According to ST 2, G_M decomposes into a union of open sets, each of which is a product over $\ell \mid M$.

It will be natural to study G by picking an integer M which splits ρ and is also stable. Then any prime $\ell \nmid M$ is also stable, or as we also say, ρ is stable at level 1 for any prime $\ell \nmid M$.

As remarked already in the introduction, Serre has shown that the torsion points of an elliptic curve over the rationals without complex multiplication give rise to a GL_2-extension as above, for which there always exists some M which splits and stabilizes ρ.

We say that the extension K, or ρ, has **limited ramification** if there is a positive integer Δ having the following property. If p is a prime and $p \nmid \Delta\ell$, then p is unramified in K_ℓ, or as we also say, ρ_ℓ is **unramified** at p. Then the Frobenius class σ_p is well defined in G_ℓ, and $\rho_\ell(\sigma_p)$ has a characteristic polynomial which we assume of the form

$$\Phi_p(X) = X^2 - t_p X + p .$$

We say that ρ is **integral and consistent** if t_p is an integer independent of ℓ. We call t_p the **trace of Frobenius**.

For the rest of this paper, we assume that ρ has limited ramification, is integral, and consistent. We assume that the roots of the characteristic polynomial are complex conjugates of each other. We say that such ρ is of elliptic type.

We let $\mathfrak{H}$ be the upper half plane, and let

$$z_p = \text{root of } \Phi_p(X) \text{ in } \mathfrak{H} .$$

Thus z_p is the representative of the Frobenius class in the upper half plane, and is also sometimes called a **Frobenius element**.

We let N denote the absolute norm to $\mathbf{Q}$, so $Nz = z\bar{z}$. If z_p is a Frobenius element, then $Nz_p = p$.

Given an integer t, a prime p, such that $t^2 < 4p$, we let

$$z(t,p) = \text{root in } \mathfrak{H} \text{ of } X^2 - tX + p = 0 .$$

We let

$$\theta(z) = \text{angle of } z \in \mathfrak{H}, \qquad 0 < \theta(z) < \pi ,$$

We sometimes write $\theta(t,p)$ for $\theta(z(t,p))$.

This settles the basic notation concerning our GL_2-extension.

§2. Some notions of probability

The law of large numbers in probability will motivate part of our work, and we state some versions of it here for the convenience of the reader.

For each positive integer n suppose given a measured space X_n with positive measure μ_n such that the total measure of X_n is 1. We let

$$X = \prod X_n \quad \text{and} \quad \mu = \prod \mu_n$$

be the product space and product measure respectively. We view X as our probability space.

Theorem 2.1. *Suppose given a measurable subset S_n of X_n for each n. Assume that the limit exists,*

$$\lim_{n \to \infty} \mu_n(S_n) = L .$$

Then for almost all elements (sequences) $x = \{x_n\}$ in X, the density of n such that $x_n \in S_n$ exists and is equal to L. This means:

$$\lim_{N \to \infty} \frac{\#\{n \leq N, \, x_n \in S_n\}}{N} = L .$$

The above theorem has a simple intuitive content, but our main application requires a stronger version as follows.

Theorem 2.2. *Suppose given a measurable subset S_n of X_n for each n. Let $\{b_n\}$ be a sequence of positive real numbers tending monotonically to infinity. Assume that*

$$\sum \frac{1}{b_n^2} \mu_n(S_n) < \infty .$$

Then for almost all sequences x, we have

$$\#\{n \leq N, \, x_n \in S_n\} = \sum_{n=1}^{N} \mu_n(S_n) + o(b_N) .$$

The first theorem is obtained from this second by putting $b_n = n$. On the other hand, we obtain the following corollary.

Corollary. *Let* p_n *be the n-th prime. Let* C *be a positive number, and assume*

$$\mu_n(S_n) \sim C \frac{1}{2\sqrt{p_n}}.$$

Then for almost all sequences x,

$$\# \{n \leq N, \, x_n \in S_n\} \sim C \sum_{n \leq N} \frac{1}{2\sqrt{p_n}}.$$

Proof. Take

$$b_N = \sum_{n \leq N} \frac{1}{2\sqrt{p_n}}.$$

The corollary is stated precisely in a form which fits the applications, where $\pi_{\frac{1}{2}}(x)$ occurs.

In the applications, the indexing set for the measured spaces will be the prime numbers, so we write p instead of n. In the simplest case, i.e. in the study of the supersingular distribution, the measured space at p will essentially be the interval of integers t such that

$$-2\sqrt{p} < t < 2\sqrt{p},$$

and the measure will be a discrete measure designed to take into account the Tchebotarev-Sato-Tate distributions. In the imaginary quadratic case, it will be similar but more complicated, and we wait till the appropriate section to give the more precise description of the probability space. It is of course a major assumption that there exists a model of the above type for the distribution of Frobenius elements under consideration. However, in the theory of prime numbers, it seems to be the appropriate model for this and other distribution problems. For instance, we can use the same pattern to recover a conjecture of Hardy-Littlewood concerning primes in quadratic progression. This will fit in a natural way in the second part.

The situation is in fact a little more complicated, because we describe a probabilistic model as above for each given positive integer M, for the sequence of Frobenius elements in the Galois group at level M, i.e. in the representation

$\rho_{(M)}$ into $GL_2(Z/MZ)$. We then have to take a further limit over M (ordered by divisibility in the supersingular case, and somewhat more subtly in the imaginary quadratic case) in order to get the asymptotic behavior for the entire Galois group.

Appendix

For the convenience of the reader we include a proof of the probabilistic theorem. The first lemma, due to Kolmogoroff and formulated by him in probabilistic terms, will be a refinement of the fundamental lemma of integration theory, which asserts that given an L^1 (or L^2) Cauchy sequence, there exists a subsequence which converges absolutely almost everywhere. Here we give up on absolute convergence, but have conditions which make the full sequence converge pointwise almost everywhere.

Lemma. *For each n let h_n be a function on X_n, also viewed as function on X by projection on the n-th factor. Assume that*

$$\int h_n \, d\mu_n = 0.$$

Let

$$H_n(x) = \sum_{k=1}^{n} h_k(x)$$

be the partial sum. Assume that $\sum \|h_k\|_2^2$ converges. Then the limit

$$\lim_{n \to \infty} H_n(x)$$

exists for almost all $x \in X$.

Proof. We first note that the functions h_n are mutually orthogonal on X. The heart of the proof lies in the next statement.

Kolmogoroff's inequality. *Given ε, let*

$$Z = \left\{ x \in X, \max_{1 \le k \le n} H_k^2(x) \ge \varepsilon \right\}.$$

Then

$$\varepsilon \mu(Z) \le \sum_{k=1}^{n} \|h_n\|_2^2.$$

Proof. Let

$$Y_k = \{x \in X, \ H_k^2(x) \geq \varepsilon \quad \text{and} \quad H_i^2(x) < \varepsilon \quad \text{all} \quad i < k\}.$$

In other words, Y_k is the set of points x such that $H_k^2(x)$ is the first partial sum at least equal to ε. Then the sets Y_k are disjoint, and we get the inequality

$$\varepsilon \sum_{k \leq n} \mu(Y_k) \leq \sum_{k \leq n} \int_{Y_k} H_k^2.$$

Write

$$H_k^2 = H_n^2 - 2H_k(H_n - H_k) - (H_n - H_k)^2.$$

The last term is negative, and we shall leave it out when we integrate. On the other hand, the middle term gives

$$\int_{Y_k} H_k(H_n - H_k) \, d\mu = 0.$$

This is due to the fact that H_k is effectively a function only of the first k variables, while $H_n - H_k$ is effectively a function only of the last $n-k$ variables. The integral splits into a product of integrals over the distinct variables, and is immediately seen to yield 0 as desired. Therefore we can replace H_k^2 by H_n^2 and then integrate over all of X, thereby giving as bound the L^2-norm squared of $\sum h_k$, which proves the asserted inequality.

We have assumed that $\sum h_k$ is in L^2, that is

$$\sum_{k=1}^{\infty} \|h_k\|_2^2 < \infty.$$

This means that for m_0 sufficiently large, and $n \geq m \geq m_0$ we get

$$\mu\{x \in X, \ \max (H_n - H_m)^2(x) \geq \varepsilon\} \leq \frac{1}{\varepsilon} \sum_{n=m}^{\infty} \|h_n\|_2^2 < \varepsilon.$$

Define

$$Z_i = \{x \in X, \ (H_n - H_m)^2(x) \geq 1/2^i \quad \text{if} \quad m, n \geq m_0(i)\}.$$

Then Z_i has measure $\leq 1/2^i$ if we pick $m_0(i)$ sufficiently large. Let

$$W_n = Z_n \cup Z_{n+1} \cup \cdots$$

for large n, so that W_n has measure $\leq 1/2^{n-1}$. Then the partial sums $\sum h_k(x)$ converge for x not in W_n. Hence if we let W be the intersection

$$W = \cap W_n ,$$

then these partial sums converge for x not in W, and W has measure zero, thereby proving the lemma.

The next theorem is also due to Kolmogoroff, in that generality.

Theorem 2.3. *For each n let f_n be a function on X_n, and assume that*

$$\int f_n \, d\mu_n = 0 .$$

Let $\{b_n\}$ be a sequence of positive real numbers monotonically increasing to infinity. If

$$\sum \frac{1}{b_n^2} \|f_n\|_2^2 < \infty ,$$

then for almost all x the partial sums

$$F_n(x) = \sum_{k=1}^{n} f_k(x)$$

satisfy the estimate

$$F_n(x) = o(b_n) .$$

Proof. Let $h_n = f_n/b_n$ and apply the lemma, to the partial sums

$$H_n(x) = \sum_{k=1}^{n} h_k(x) = \sum_{k=1}^{n} f_k(x)/b_k .$$

The lemma says that these partial sums converge for almost all x. It is a trivial fact (proved by summation by parts) that if $\sum a_k$ is a convergent sequence then

$$\sum_{k=1}^{n} a_k b_k = o(b_n).$$

Applying this fact when $a_k = h_k(x)$ proves the theorem.

We have stated Theorem 2.3 under the normalization that the integral of the functions f_n is 0. This is of course not satisfied in general, but a translation reduces the general case to this special case. Indeed, suppose that ψ_n are functions such that

$$\int \psi_n \, d\mu_n = c_n$$

is a constant c_n. Define

$$f_n = \psi_n - c_n.$$

Then the integral of f_n is 0. In particular, suppose that ψ_n is the characteristic function of some subset S_n of X_n. Then

$$\|f_n\|_2^2 = \int (\psi_n - c_n)^2 \, d\mu_n = c_n - c_n^2.$$

Applying Theorem 2.3 to this situation yields Theorem 2.2, as desired.

THE DISTRIBUTION FOR FIXED TRACE

§3. The probabilistic model

We fix a positive integer M. For convenience, instead of an interval, we let the measured space associated with each prime p be the set of integers Z. The measure μ_p on each fiber is assumed to be represented by a function

$$f_M(t, p) \geq 0, \qquad t \in Z,$$

with respect to counting measure. We also write

$$f_M(t, p) = \text{pr}\{y_p = t\},$$

where y_p is the "random variable" in the language of the probabilists. The probability condition is that

$$\sum_t f_M(t, p) = 1,$$

and we assume in addition that

PR 1. $\qquad f_M(t, p) = 0 \quad if \quad |t| > 2\sqrt{p}.$

Let S be a congruence class mod M, say consisting of all those integers $\equiv t \pmod{M}$. Let $G(M)_t$ be defined by

$$G(M)_t = \{\sigma \in G(M), \text{ tr } \sigma \equiv t \bmod M\}.$$

By Tchebotarev, the density of primes p such that $t_p \equiv t \pmod{M}$ has a density equal to

$$\frac{|G(M)_t|}{|G(M)|}.$$

We define

$$F_M(t) = M \frac{|G(M)_t|}{|G(M)|},$$

so F_M depends only on the residue class of t mod M. The factor M is put there so that the average value of F_M is 1, i.e.

$$\sum_{t \bmod M} \tfrac{1}{M} F_M(t) = 1 .$$

For real y, define
$$\xi(y,p) = \frac{y}{2\sqrt{p}} ,$$
so that if $|\xi(y,p)| < 1$, then $\xi(y,p) = \cos\theta(y,p)$. We assume

ST. *There exists a continuous density function $\phi(\theta)$ on $[0,\pi]$ determining the distribution of angles $\theta(t_p, p)$.*

This means that the primes p such that $\theta(t_p, p)$ lies in a given interval $[\theta_1, \theta_2]$ have a density, namely
$$\int_{\theta_1}^{\theta_2} \phi(\theta)\, d\theta .$$

Changing variables, with $\xi = \cos\theta$ shows that the condition ST is equivalent with:

ST'. *There exists a continuous function*
$$g(\xi) = \frac{\phi(\theta)}{\sin\theta}$$
which is equal to 0 outside the interval $[-1, 1]$, such that the primes p for which $\xi(t_p, p)$ lies in an interval $[\xi_1, \xi_2]$ have a density, given by
$$\int_{\xi_1}^{\xi_2} g(\xi)\, d\xi .$$

We call g (or ϕ) the distribution function of Frobenius at infinity. Note that $g = g_\rho$ depends on ρ. When the representation ρ is that obtained from the torsion points of an elliptic curve without complex multiplication, then the Sato-Tate conjecture states that
$$\phi(\theta) = \tfrac{2}{\pi} \sin^2\theta .$$

The function ϕ is of course normalized to have integral 1 over $[0, \pi]$. In this special case, we have
$$g(\xi) = \tfrac{2}{\pi} \sqrt{1-\xi^2} .$$

Our main assumption concerning the probabilistic model is that the function f_M can be written in the form

PR 2. $$f_M(t, p) = c_p g(\xi(t, p)) F_M(t)$$

where c_p is a constant which is determined so that

$$\sum_t f_M(t, p) = 1.$$

This is an assumption of independence for the behavior at infinity and over congruence classes. We shall now determine the asymptotic nature of c_p, and prove that

$$c_p \sim \frac{1}{2\sqrt{p}}.$$

The integral

$$1 = \int_{-1}^{1} g(\xi) \, d\xi$$

has approximating Riemann sums

$$\frac{1}{2\sqrt{p}} \sum_{-2\sqrt{p} < t < 2\sqrt{p}} g(\xi(t, p)), \qquad \text{for } p \to \infty.$$

For any integer t_0 we have the limit

(1) $$\lim_p \frac{1}{2\sqrt{p}} \sum_{t \equiv t_0 \bmod M} g(\xi(t, p)) = \frac{1}{M}.$$

Hence for given M we have the limit as $p \to \infty$,

(2) $$\lim_p \frac{1}{2\sqrt{p}} \sum_{t \equiv t_0 (M)} g(\xi(t, p)) F_M(t_0) = \frac{F_M(t_0)}{M}.$$

Summing over the congruence classes t_0, we find that

$$\lim_p \frac{1}{2\sqrt{p}} \sum_t g(\xi(t, p)) F_M(t) = 1.$$

Comparing with the definition and normalization of $f_M(t, p)$ proves that

$$c_p \sim \frac{1}{2\sqrt{p}},$$

as desired.

Going back to (2) with the knowledge that $c_p \sim \frac{1}{2\sqrt{p}}$ now shows:

Lemma 1. $$\lim_p \mu_p(S) = \frac{1}{M} F_M(t_0)$$

where S is the congruence class of t_0 mod M.

The law of large numbers then proves that our probability assumption PR 2 implies that the density of p such that

$$y_p \equiv t_0 \bmod M$$

is precisely equal to

$$\frac{1}{M} F_M(t_0),$$

i.e. that our probability assumption is compatible with the Tchebotarev density property for the sequence of Frobenius elements to be viewed as a random sequence in our model.

Given a subinterval I of $[-1, 1]$, let $S_{I,p}$ be the set of integers y_p such that $\xi(y_p, p) \in I$. By definition,

$$\mu_p(S_{I,p}) = \sum_{\xi(t,p) \in I} f_M(t, p).$$

Lemma 2. $$\lim_p \mu_p(S_{I,p}) = \int_I g(\xi)\, d\xi.$$

Proof. Entirely analogous to the proof of Lemma 1. We start with approximating Riemann sums for the integral

$$\int_I g(\xi)\, d\xi,$$

and use the analogue of (1) and (2) which have this number as a factor instead of the number 1. We leave the details to the reader.

Now the law of large numbers yields the compatibility with the Sato-Tate conjecture.

§4. The asymptotic behavior

Given the probabilistic model of the last section, and a random sequence $\{y_p\}$, and noting that

$$\xi(t_0, p) \longrightarrow \xi(0) = \phi(\pi/2) \qquad \text{as } p \longrightarrow \infty,$$

we get

$$\mathrm{pr}_M\{y_p = t_0\} = f_M(t_0, p) = c_p g(\xi(t_0, p)) F_M(t_0)$$

$$\sim \frac{1}{2\sqrt{p}} g(0) F_M(t_0)$$

$$\sim \frac{1}{2\sqrt{p}} \phi(\pi/2) F_M(t_0).$$

It is now reasonable to suppose that taking the limit for $M \to \infty$ (under the ordering by divisibility) yields the probability that $t_p = t_0$, or in other words, that the above value for large M gives an approximation of this probability. In fact, it will be proved in Theorem 4.2 that the limit

$$\lim_M F_M(t_0) = F(t_0)$$

exists, and is given by an absolutely convergent product. We then define

$$\boxed{C(t_0, \rho) = \phi(\pi/2) F(t_0)}$$

and our conjecture is that the number of primes $p \leq x$ such that $t_p = t_0$ satisfies the asymptotic property

$$\boxed{N_{t_0}(x) \sim C(t_0, \rho) \pi_{\frac{1}{2}}(x).}$$

In the case of elliptic curves, $\phi(\pi/2) = 2/\pi$.

The factor $F(t_0)$ arose from congruence conditions, and we now see how it decomposes into a product over primes.

Recall that
$$F_M(t) = M \frac{|G(M)_t|}{|G(M)|}.$$

We proceed to determine F_M as much as we can in a general context. First we have the multiplicativity.

Lemma 1. *Suppose that* $M = M_1 M_2$ *is a product of two relatively prime factors, and* $G(M) = G(M_1) \times G(M_2)$. *Then*
$$F_M(t_0) = F_{M_1}(t_0) F_{M_2}(t_0).$$

Proof. Obvious.

Next we study F_M as M becomes highly divisible. We first show that the limit $\lim_{n \to \infty} F_{M^n}(t)$ depends only on a stabilizing integer M_0, and then find an explicit value for the prime power factor for almost all ℓ. As a matter of notation, we write
$$F_{M^\infty}(t) = \lim_{n \to \infty} F_{M^n}(t)$$
whenever that limit exists. We identify $G(M)$ inside $GL_2(Z(M)) = GL_2(M)$.

Lemma 2. *Assume that* M_0 *is stable, that* $M_0 | M$, *and that* M_0 *is divisible by the same primes as* M. *If* $s \equiv t \pmod{M_0}$ *then*
$$|G(M)_s| = |G(M)_t|.$$
Furthermore,
$$F_M(t) = F_{M_0}(t).$$

Proof. There is a bijection between the two sets $G(M)_t$ and $G(M)_s$, given by
$$\sigma \longmapsto \sigma + \begin{pmatrix} s-t & 0 \\ 0 & 0 \end{pmatrix}$$
which makes the first assertion obvious. If $r: GL_2(M) \to GL_2(M_0)$ is the reduction map, then given $\bar{\sigma} \in GL_2(M_0)$ there are $(M/M_0)^4$ elements in the fiber $r^{-1}(\bar{\sigma})$, and there are M/M_0 elements of Z/MZ lying above an element of $Z/M_0 Z$. Hence by the first part of the lemma, we find

$$(M/M_0)^4 |G(M_0)_t| = (M/M_0) |G(M)_t|,$$

whence

$$M |G(M)_t| = (M/M_0)^4 M_0 |G(M_0)_t|.$$

But

$$|G(M)| = (M/M_0)^4 |G(M_0)|.$$

Taking the quotient proves the lemma.

In particular, *if* M *is stable, then*

$$F_{M^n}(t) = F_M(t)$$

is constant, so the limit $F_{M^\infty}(t)$ *obviously exists and is equal to* $F_M(t)$.

Lemma 3. *If* $uI \in G(M)$ *for some integer* u (mod M), *then*

$$F_M(ut) = F_M(t).$$

Proof. The map $\sigma \mapsto u\sigma$ is a bijection of $G(M)_t$ and $G(M)_{ut}$.

Theorem 4.1. *Assume that* $G_\ell = GL_2(Z_\ell)$. *Let* $r = r(\ell) = 1/\ell$. *Then*

$$F_{\ell^\infty}(0) = F_\ell(0) = \frac{1}{1-r^2}$$

$$F_{\ell^\infty}(1) = F_\ell(1) = \left(1 - \frac{r^3}{1-r-r^2}\right)^{-1}.$$

Proof. By Lemma 3, and Lemma 2, we only have to consider $F_\ell(0)$ and $F_\ell(1)$, i.e. $t = 0$ or $t = 1$ at level 1.

Suppose first $t = 0$. Then $G(\ell)_0$ consists of the matrices

$$\begin{pmatrix} a & b \\ c & -a \end{pmatrix} \text{ such that } a^2 + bc \not\equiv 0 \bmod \ell.$$

If $a \equiv 0$, there are $(\ell-1)^2$ pairs b, c which give rise to possible matrices. the other hand, there are $\ell-1$ values for $a \not\equiv 0$. Given b, there is a unique such that $bc \equiv -a^2$, so there are

$$\ell^2 - (\ell - 1)$$

pairs b, c such that $a^2 + bc \not\equiv 0 \pmod{\ell}$. Adding yields

$$|G(\ell)_0| = (\ell-1)^2 + (\ell-1)(\ell^2 - (\ell-1)) = \ell^2(\ell-1).$$

Furthermore, to get the case $t = 1$, we use the formula

$$\sum_{t \bmod \ell} |G(\ell)_t| = |G(\ell)_0| + (\ell - 1)|G(\ell)_1|$$

and also

$$\sum_{t \bmod \ell} |G(\ell)_t| = |GL_2(F_\ell)| = \ell(\ell-1)(\ell^2 - 1).$$

We then solve for $|G(\ell)_1|$, to get

$$|G(\ell)_1| = \ell(\ell^2 - \ell - 1).$$

Thus

$$F_\ell(0) = \frac{\ell |G(\ell)_0|}{|GL_2(F_\ell)|} = \frac{\ell^2}{\ell^2 - 1} \quad \text{and} \quad F_\ell(1) = \frac{\ell(\ell^2 - \ell - 1)}{(\ell-1)(\ell^2 - 1)},$$

as desired.

We can put all these computations together. Using the multiplicativity lemma and the stabilizing Lemma 2, we find:

Theorem 4.2. *Assume that M splits and stabilizes ρ. Then the constant is given by*

$$C(t_0, \rho) = \phi(\pi/2) F_M(t_0) \prod_{\ell \nmid M} F_\ell(t_0).$$

In particular,

$$C(0, \rho) = \phi(\pi/2) \frac{\pi^2}{6} F_M(0) \prod_{\ell \mid M} (1 - 1/\ell^2).$$

Proof. The first assertion follows by putting the lemmas together. We then use the value found in Theorem 4.1,

$$F_\ell(0) = \frac{1}{1 - 1/\ell^2}.$$

Since $\zeta(2) = \pi^2/6$, the second formula follows at once.

In this expression, it is the factor $F_{M^\infty}(0)$, depending on the curve A, which distinguishes the special situation. This factor is entirely determined by the image of the representation

$$\rho_{(M)} : G(M) \longrightarrow GL_2(Z(M)) .$$

In the case of an elliptic curve, it follows from results of Serre (cf. the next section) that any M which splits and stabilizes ρ is necessarily even.

Remark 1. *If $t = 0$ then the constant $C(0, \rho)$ is not 0.* To see this, it suffices to observe that there is at least one element in the stable group $G(M)$ with trace 0. Such an element is given by complex conjugation, for instance, so that $F_M(0) \neq 0$ for a stabilizing M.

On the other hand, for other values of $t_0 \neq 0$ it may happen that the constant $C(t_0, \rho)$ is equal to 0. A typical reason for this is the presence of rational points on the elliptic curve, which may impose congruence conditions on the traces of Frobenius.

Remark 2. Let A, B be two elliptic curves over the rationals, without complex multiplication. Then the Galois group $\mathcal{G}$ of $Q(A_{tor}, B_{tor})$ over the rationals is contained in the subgroup of the product of the Galois groups of $Q(A_{tor})$ and $Q(B_{tor})$ consisting of the elements which have the same effect on the field of all roots of unity, and Serre conjectured that it is of finite index (he also proved this when one of the j-invariants is not integral). Assuming that this is the case, and arguing heuristically in a manner analogous to that used for one curve, assuming also in addition that probability distributions on the two curves are essentially independent, one comes up with an estimated asymptotic behavior for the set of primes p which are supersingular for both A and B. The number of such primes $\leq x$ should in particular be

$$O\left(\sum_{p \leq x} \frac{1}{p}\right) = O(\log \log x) .$$

Therefore, the conjecture concerning a strengthening of the isogeny theorem is that if two elliptic curves over the rationals without complex multiplication have a common set of supersingular primes of asymptotic order of magnitude which is not $O(\log \log x)$, then the curves are isogenous.

For the curves we looked at, there are just two coincidences of supersingular primes among the first 5,000 primes. The prime 19 is supersingular for A and $X_0(11)$, and 2,411 is supersingular for C and D. The function $\log \log x$ grows so slowly that it would be hard to get statistically meaningful data.

Of course a similar conjecture can be made about primes whose Frobenius elements for two given curves both generate the same quadratic field.

EXAMPLES

§5. Serre curves, $M = 2q$, the general formula

Serre has shown that G is always contained in a subgroup of index 2 in $\prod GL_2(Z_\ell)$, determined by conditions on the quadratic symbol with respect to the field $Q(\sqrt{\Delta})$. For our purposes, we need only a special case.

Let q be an odd prime. The group $GL_2(Z_q)$ has a unique subgroup of index 2, denoted by E_q. It is the set of elements σ such that $\det \sigma$ is a square in Z_q^* (or equivalently, a square mod q). Since the determinant of the scalar matrices consists of the squares, we see that E_q is simply the product of the scalar matrices times $SL_2(Z_q)$.

For the prime 2, we have the reduction homomorphism

$$GL_2(Z_2) \longrightarrow GL_2(2) \approx S_3 ,$$

and we let E_2 be the subgroup of elements σ such that the reduction $\bar\sigma$ mod 2 is an even permutation.

We call E_q or E_2 the group of even elements, and let O_q, O_2 be their respective cosets, called the cosets of odd elements. We define Serre's subgroup S_{2q} of $GL_2(Z_2) \times GL_2(Z_q)$ to be

$$S_{2q} = (E_2 \times E_q) \cup (O_2 \times O_q) .$$

It is of index 2, and $2q$ stabilizes this subgroup. Cf. [S 2], 5.5, Proposition 22, where Serre shows that if the discriminant of the field $Q(\sqrt{\Delta})$ is $\pm q^n$, then the image of ρ is always contained in

$$S_{2q} \times \prod_{\ell \neq 2, q} GL_2(Z_\ell) ,$$

which we call Serre's subgroup of $\prod GL_2(Z_\ell)$.

If $M = 2q$ is stable, and $G(M) = S_{2q}(M)$, we also call $G(M)$ the Serre subgroup in $GL_2(Z(M))$.

As the supersingular case is especially important, the next two theorems give the value of $F_M(0)$. Complete tables are established after that for all values $F_M(t)$.

Theorem 5.1. *Let* $M = 2q$ *with* q *an odd prime. Assume:*
(i) M *is stable.*
(ii) $G(M)$ *is Serre's subgroup of index* 2 *in* $GL_2(Z(M))$.
Then

$$F_M(0) = \begin{cases} \dfrac{2q(2q^2 - 3q + 1)}{3(q-1)(q^2-1)} & \text{if } \left(\dfrac{-1}{q}\right) = 1 \\[2ex] \dfrac{2q(2q^2 - q - 1)}{3(q-1)(q^2-1)} & \text{if } \left(\dfrac{-1}{q}\right) = -1 . \end{cases}$$

Remark. The two cases depending on the parity of $\left(\dfrac{-1}{q}\right)$ have values which differ only in a second order term, and tend to $\dfrac{4}{3}$ as $q \to \infty$.

Theorem 5.1 will be proved below. If we combine the values given for $F_{2q}(0)$ in this theorem with the general expression for the asymptotic supersingular distribution found in §4, we get the asymptotic supersingular distribution in the Serre case.

Theorem 5.2. *Let* $M = 2q$ *where* q *is an odd prime. Assume:*
(i) M *splits and stabilizes* ρ.
(ii) $G(M)$ *is Serre's subgroup of index* 2 *in* $GL_2(Z(M))$.
Then

$$C(0, \rho) = \phi\left(\frac{\pi}{2}\right) \frac{\pi^2}{6} F_{2q}(0)(1 - 1/4)(1 - 1/q^2) .$$

In particular, for the ordinary Sato-Tate distribution,

$$C(0, \rho) = \frac{\pi}{4} F_{2q}(0)(1 - 1/q^2) .$$

We shall now compute the values $F_{2q}(t)$ when q is an odd prime and $2q$ is stable. By definition,

$$F_{2q}(t) = \frac{2q |G(2q)_t|}{3q(q-1)(q^2-1)} ,$$

because $G(2q)$ is of index 2 in

$$GL_2(2) \times GL_2(q) ,$$

and $GL_2(2)$ has order 6, while $GL_2(q)$ has order $q(q-1)(q^2-1)$. Thus our task is to compute $G(2q)_t$ for various values of t. The result is given in the following table. Since it turns out that $|G(2q)_t|$ is divisible by q in all cases, the table gives the quotient by q, i.e. it gives the values of

$$\tfrac{1}{q} |G(2q)_t| .$$

		$q \mid t$	$q \nmid t$
$\left(\tfrac{-1}{q}\right) = 1$	$2 \mid t$	$2q^2 - 3q + 1$	$2q^2 - 2q - 1$
	$2 \nmid t$	$q^2 - 1$	$q^2 - q - 2$
$\left(\tfrac{-1}{q}\right) = -1$	$2 \mid t$	$2q^2 - q - 1$	$2q^2 - 2q - 3$
	$2 \nmid t$	$q^2 - 2q + 1$	$q^2 - q$

For example, if $\left(\tfrac{-1}{q}\right) = 1$, then the table shows that

$$|G(2q)_0| = q(2q^2 - 3q + 1)$$

and consequently $F_{2q}(0)$ has the value stated in Theorem 5.1. In general, we may write

$$F_{2q}(t) = \frac{2q^2 \cdot \text{table entry}}{3q(q-1)(q^2-1)} .$$

We now give the proof that the table is correct.

By definition, Serre's subgroup G is the group

$$G = [(E_2 \times E_q) \cup (O_2 \times O_q)] \times \prod_{\ell \nmid 2q} GL_2(Z_\ell) .$$

Our notation is such that

$$E(q)_t = \{\sigma \in E(q), \text{ tr } \sigma \equiv t \bmod q\} .$$

We note that if $u \in F_q^*$ and $u \neq 0$, then the map

$$\sigma \longmapsto u\sigma$$

gives a bijection

$$E(q)_t \longrightarrow E(q)_{ut},$$

so the sets $E(q)_t$ all have the same cardinality for $t \neq 0$. Hence

$$|E(q)| = \tfrac{1}{2} q(q-1)(q^2-1) = |E(q)_0| + (q-1)|E(q)_1|.$$

We compute $|E(q)_0|$. We note that there is a direct product decomposition

$$E(q)_0 = (F_q^*/\pm 1) SL_2(q)_0,$$

where $F_q^*/\pm 1$ denotes a set of representatives of F_q^* mod ± 1. Thus it suffices to count the order of $SL_2(q)_0$. Let

$$\sigma = \begin{pmatrix} a & b \\ c & -a \end{pmatrix}$$

be an element of $SL_2(q)_0$, so $-bc = 1+a^2$. We want to know how many such elements there are.

Case 1. $\left(\frac{-1}{q}\right) = 1$.

There are two values of a such that $1+a^2 = 0$, and to each such value there are $2q-1$ pairs (b,c) with $-bc = 0$. There are $q-2$ values of a such that $1+a^2 \neq 0$, and for each such a, there are $q-1$ pairs (b,c) with $-bc = 1+a^2$. This yields

$$|SL_2(q)_0| = 2(2q-1) + (q-1)(q-2) = q^2 + q$$

whence

$$|E(q)_0| = \tfrac{1}{2}(q^3 - q).$$

Case 2. $\left(\frac{-1}{q}\right) = -1$.

Then $1+a^2 \neq 0$ for all a, whence

$$|SL_2(q)_0| = q^2 - q \quad \text{and} \quad |E(q)_0| = \tfrac{1}{2}(q^3 - 2q^2 + q).$$

Next, we have
$$GL_2(q)_0 = E(q)_0 \cup O(q)_0$$
from which we can compute the order of $O(q)_0$. Arithmetic then yields the following table of orders for $E(q)_t$ and $O(q)_t$.

	$t = 0$	$t \neq 0$
$\left(\frac{-1}{q}\right) = 1$, $E(q)_t$	$\frac{1}{2} q(q^2 - 1)$	$\frac{1}{2} q(q^2 - q - 2)$
$\left(\frac{-1}{q}\right) = -1$, $O(q)_t$		
$\left(\frac{-1}{q}\right) = 1$, $O(q)_t$	$\frac{1}{2} q(q^2 - 2q + 1)$	$\frac{1}{2} q(q^2 - q)$
$\left(\frac{-1}{q}\right) = -1$, $E(q)_t$		

The individual case for the prime 2 is done by inspection:

	$t = 0$	$t = 1$
$E(2)_t$	1	2
$O(2)_t$	3	0

The table for $|G(2q)_t|$ then follows trivially.

In order to apply these tables to concrete instances, we must determine explicitly that certain elliptic curves have the Serre group as Galois group of their division points, or groups related to this subgroup. The next sections are devoted to the proofs in two important instances which are fairly typical, and relatively complicated.

§6. Computations of Galois groups

In this section we recall some techniques for the computation of Galois groups.

Refinement lemmas

The first lemma was proved originally by Shimura [Sh] with some restrictive conditions, and was proved by Serre [S 3] in general. See also [L 1], Chapter 17, §4

Lemma 1. *Let* H *be a closed subgroup of* $GL_2(Z_\ell)$ *whose projection mod* ℓ *contains* $SL_2(F_\ell)$. *Then* H *contains* $SL_2(Z_\ell)$ *if* $\ell \geq 5$.

We refer to this lemma as the refinement lemma. There are others of a slightly more abelian nature as follows. Cf. Shimura [Sh].

Let $M_\ell = Mat_2(Z_\ell)$, and since ℓ is fixed, write $M = M_\ell$. Let

$$V_n = I + \ell^n M_\ell$$

be the group of matrices $\equiv I \mod \ell^n$. The map

$$X \longmapsto I + \ell^n X$$

induces an isomorphism

$$M/\ell M \xrightarrow{\approx} V_n/V_{n+1} .$$

Consider the case when $\ell \geq 3$. The map raising to the ℓ-th power gives an isomorphism

$$V_n/V_{n+1} \approx V_{n+1}/V_{n+2} .$$

Let U_1 be a closed subgroup of V_1 and let $U_n = U_1 \cap V_n$. We have an embedding

$$U_1/U_2 \longrightarrow V_1/V_2 \approx M/\ell M .$$

We may refer to U_1/U_2 as the associated vector space to U_1 at level 1, or as the tangent space to U_1. It has dimension ≤ 4.

Lemma 2. *Let ℓ be odd, and let U_1 be a closed subgroup of V_1 whose associated vector space at level 1 has dimension 4. Then $U_1 = V_1$.*

Proof. From the isomorphism $U_1/U_2 \approx V_1/V_2$ raised to the ℓ-th power recursively, we get an isomorphism

$$U_n/U_{n+1} \approx V_n/V_{n+1},$$

and it then follows that $U_1/U_n = V_1/V_n$, whence taking the limit shows $U_1 = V_1$, as desired.

For the prime 2, as usual, one has to start the recursive process mod 8, so we have an isomorphism

$$V_2/V_3 \approx V_n/V_{n+1}, \qquad n \geq 2.$$

From this we get the similar lemma:

Lemma 3. *Let $\ell = 2$. Let U_1 be a closed subgroup of V_1. If $U_2/U_3 = V_2/V_3$ then $U_2 = V_2$. If furthermore $U_1/U_3 = V_1/V_3$, in other words, if the reduction of U_1 mod 8 contains*

$$I + 2M_2 \pmod{8},$$

then $U_1 = V_1$.

Proof. Clear.

Abelianization

We let G be our usual group with representation ρ. The field K of which G is the Galois group contains all the roots of unity, and this cyclotomic field is the maximal abelian extension of the rationals.

Let G' denote the closure of the commutator subgroup, and $G^{ab} = G/G'$. Then G^{ab} is the Galois group of the cyclotomic field.

Lemma 4. *Assume that M is divisible by all the primes dividing Δ. Then*

$$G'_M = G_M \cap \prod_{\ell \mid M} SL_2(\mathbb{Z}_\ell).$$

Proof. We observe that SL_2 is the kernel of the det map,

$$\det : \prod GL_2(Z_\ell) \longrightarrow \prod Z_\ell^* \; .$$

By Kronecker's theorem, the only abelian extensions of the rationals ramified only at primes dividing M are contained in the field of M^∞-th roots of unity, and the lemma follows at once.

Remark. If no assumption is made concerning the ramification, then one has only an inclusion

$$G'_M \subset G_M \cap \prod_{\ell | M} SL_2(Z_\ell) \; .$$

For an example of this which causes trouble, see the section on the Shimura curve

A theorem of Serre

In this part we recall techniques of Serre showing that the Galois group of division points is nearly GL_2.

Let G be a subgroup of a product

$$G \subset (G_1 \times G_2) \; ,$$

and assume that the projections of G on the two factors are G_1 and G_2 respectively. Let

$$H_1 = \text{pr}_1 [G \cap (G_1 \times \{e_2\})] \quad \text{and} \quad H_2 = \text{pr}_2 [G \cap (\{e_1\} \times G_2)] \; .$$

Then there is an isomorphism

$$G_1/H_1 \approx G_2/H_2$$

whose graph is induced by G. We call this **Goursat's lemma**.

If L is a possibly infinite set of primes, and G is again the Galois group of the division points of an elliptic curve A over Q, we let G_L be the projection on the L-th factor. In other words, we have the representation

$$\rho_L : G \longrightarrow \prod_{\ell \in L} GL_2(Z_\ell) \; ,$$

and we denote its image by G_L.

We say that a prime occurs in a group if it divides the order of some solvable Jordan-Holder constituent for the group.

We let $M_\ell = \text{Mat}_2(Z_\ell)$.

Theorem 6.1. *Let M be divisible by 2, 3 and all primes dividing the discriminant Δ of A. Let L consist of the primes not dividing M. Assume:*

(i) $G(\ell) = GL_2(F_\ell)$ *for all* $\ell \in L$.

(ii) *If* $\ell \in L$, *then* ℓ *does not occur in* G_M.

Then M splits ρ, and
$$G_L = \prod_{\ell \in L} GL_2(Z_\ell).$$

Proof. This is proved by Serre's arguments reproduced in [L 1], Chapter 17, §4 and §5. We sketch the proof. We have $G \subset G_M \times G_L$. We let H_M and H_L be the corresponding Goursat subgroups. The refinement lemma, together with the fact that
$$SL_2(F_\ell)/\pm 1$$
is simple, imply that the groups
$$(\cdots, 1, SL_2(Z_\ell), 1, \cdots), \qquad \ell \in L,$$
are contained in G_L and also in H_L. Since $I + \ell M_\ell$ is an ℓ-group, the hypothesis (ii) implies also that
$$(\cdots, 1, I + \ell M_\ell, 1, \cdots)$$
is contained in H_L. Using the determinant map which gives the effect on the field of roots of unity then shows that G_L is the full product of all $GL_2(Z_\ell)$, and G_L/H_L is abelian. It is the Galois group of a cyclotomic field which ramifies only in L, and whose intersection with the field of G_M/H_M must be Q, because only primes dividing M ramify in ρ_M. Hence H_L is all of G_L, and it follows that M splits ρ, thus proving the theorem.

The techniques involved in determining certain Galois groups involve all the above, and two main additional ones:

First the use of the Tate parametrization, which locally gives unipotent elements, cf. [L 1], Chapter 15.

Second, the use of Frobenius elements, whose traces are known from tables, and which we find ad hoc to satisfy certain congruence properties, ultimately showing that the Galois group is large. This is related to the method used by Shimura [Sh] when he investigated the Galois group of division points of $X_0(11)$.

§7. The curve $y^2 = x^3 + 6x - 2$

The above curve is considered by Serre as 5.9.2, p. 318, [S 2]. It has

$$\Delta = -2^6 3^5 \quad \text{and} \quad j = 2^9 3.$$

Serre has shown that $G_\ell = GL_2(Z_\ell)$ for $\ell \neq 2, 3$, and also that

$$G(2) = GL_2(F_2) \quad \text{and} \quad G(3) = GL_2(F_3).$$

Theorem 6.1 shows that 6 splits ρ. All we have to determine is

$$\rho_{2,3} : G \longrightarrow G_2 \times G_3,$$

and we shall prove:

Theorem 7.1. *The above curve is a Serre curve with* $q = 3$.

The proof is given in the following lemmas.

Lemma 1. $G(6) = S_6(6)$.

Proof. Compatibility on the field $Q(\sqrt{\Delta})$ requires that the group of torsion points is contained in the Serre group, so we have $G(6) \subset S_6(6)$. A machine computation checks that the points of order 2 have degree 6 over Q, and that the cubic field above the quadratic field $Q(\sqrt{\Delta})$ is not contained in the field of 3-torsion points. Indeed, the machine finds that the prime 67 splits completely in the field of x-coordinates of A_3 (A is the curve), but does not split completely in $Q(A_2)$. The equality $G(6) = S_6(6)$ follows because the orders of the two groups are equal.

The image $\rho_{2,3}(G)$ in $G_2 \times G_3$ gives a Goursat correspondence, and we let H_2, H_3 be the corresponding kernels, so that

$$G_2/H_2 \approx G_3/H_3.$$

We must determine H_2 and H_3.

Lemma 2. $H_3 \supset I + 3M_3$ and $G_3 = GL_2(Z_3)$.

Proof. For the prime 23, tables show that $t_{23} = 6$. The characteristic polynomial of Frobenius σ_{23} is

$$X^2 - 6X + 23 \equiv (X-1)(X+4) \pmod{9}.$$

Since $-1 \not\equiv 4 \pmod{3}$ we can diagonalize $\rho_3(\sigma_{23})$ over Z_3, and

$$\rho_3(\sigma_{23}) \equiv \begin{pmatrix} 1 & 0 \\ 0 & -4 \end{pmatrix} \pmod{9}.$$

Furthermore,

$$\rho_3(\sigma_{23}^2) \equiv I + 3 \begin{pmatrix} 0 & 0 \\ 0 & 5 \end{pmatrix} \pmod{9}.$$

But $\left(\frac{23}{3}\right) = -1$, and $\rho_2(\sigma_{23})$ is odd in G_2, so

$$\rho_2(\sigma_{23}^2) \equiv I \pmod{2}.$$

Taking 4-th powers, we conclude:
The sequence $\rho_2(\sigma_{23})^{4n}$ converges to I in G_2.
A subsequence of $\rho_3(\sigma_{23})^{4n}$ converges to an element r in G_3, and

$$r \equiv I + 3 \begin{pmatrix} 0 & 0 \\ 0 & 5 \end{pmatrix} \pmod{9}.$$

Since $\rho_2(r) = I$, it follows that r lies in H_3. We write

$$r = I + 3Y.$$

Since H_3 is normal in G_3, the elements

$$I + 3g\,Y\,g^{-1} \pmod{9}$$

are also in H_3 for $g \in GL_2(3)$. Hence the associated vector space in $M_3/3M_3$ is 4-dimensional. This shows that H_3 contains $I + 3M_3$. Since Serre proved that $G(3) = GL_2(F_3)$, it follows also that $G_3 = GL_2(Z_3)$, thereby proving the lemma.

Lemma 3. $H_2 \supset I + 2M_2$ and $G_2 = GL_2(Z_2)$.

Proof. Since $\left(\frac{31}{3}\right) = 1$, it follows that the Frobenius element σ_{31} is even in $G(2)$. Tables show that $t_{31} = 8$, so

$$\sigma_{31} \in I + 2M_2 .$$

Write $\sigma_{31} = I + 2Y$. Then the characteristic equation for σ_{31} is

$$X^2 - 8X + 31 = 0 ,$$

whence the characteristic polynomial for Y is

$$Y^2 - 3Y + 6 \equiv (Y+2)(Y+3) \pmod{8} .$$

Since $2 \not\equiv 3 \pmod{2}$ we can diagonalize Y over Z_2. Hence

$$\rho_2(\sigma_{31}) \equiv I + 2 \begin{pmatrix} -2 & 0 \\ 0 & -3 \end{pmatrix} \pmod{8} .$$

Furthermore, $\rho_3(\sigma_{31})$ is a root of

$$X^2 - 8X + 31 \equiv X^2 + X + 1 \pmod{3} .$$

Hence

$$\rho_3(\sigma_{31})^3 \equiv I \pmod{3} .$$

It follows that the sequence $(\sigma_{31}^3)^{9^n}$ converges to I in G_3. There is a subsequence which converges to an element τ in G_2, and τ must lie in H_2 since its projection by ρ_3 is trivial. We must have

$$\tau \equiv I + 2 \begin{pmatrix} -2 & 0 \\ 0 & -3 \end{pmatrix} \pmod{8} .$$

Conjugation by elements of $GL_2(2)$ shows that we get all elements of

$$I + 2M_2 \pmod{4}$$

in $H_2 \pmod{4}$. Hence

$$G(4) = GL_2(4) .$$

Once more, conjugation by $GL_2(4)$ shows that we get all elements of

$$I + 2M_2 \pmod{8}$$

in H_2 (mod 8). The recursive procedure takes hold, and we have found that H_2 contains $I + 2M_2$. Using Serre's result that $G(2) = GL_2(F_2)$, we conclude that $G_2 = GL_2(Z_2)$, thereby proving the lemma.

The theorem is immediate from the lemmas.

Remark. The use of Frobenius elements in the above proof and in the next section is similar to that made by Shimura [Sh].

§8. The Shimura curve $X_0(11)$

The curve $X_0(11)$ is defined over the rationals by the equation

$$y^2 + y = x^3 - x^2 - 10x - 20.$$

It has $\Delta = -11^5$ and $j = -2^{12}\,31^3/11^5$. Cf. Shimura [Sh]. We let ρ be its Galois representation, and G as before the Galois group of all its division points. A number of facts are known about G, and we shall recall the proofs of some of them, as we determine ρ.

The crucial primes are 2, 5, 11. We begin by looking at 5. The equation satisfied by the x-coordinate of the points of order 5 (actually the equation for the points of order n, given recursively) can be found in Weber, among other places. In the present case, it is determined by machine computation to be

$$0 = (x-5)(x-16)(5x^2+5x-29)(x^4+15x^3+120x^2+200x+155)(x^4+x^3+11x^2+41x+101).$$

The linear factor shows that there is a rational point of order 5, whose coordinates are given by $(5,5)$. Actually, one can even exhibit the coordinates of its integral multiples, namely
$$(5,5),\ (5,-6),\ (16,60),\ (16,-61).$$

In particular, $G(5)$ is contained in the upper triangular Borel subgroup. The fact that the above equation has no irreducible factor of degree ≥ 5 shows that the group $G(5)$ cannot contain an element of order 5. Therefore we conclude that

$$G(5) = \left\{ \begin{pmatrix} 1 & 0 \\ 0 & u \end{pmatrix},\ u \in F_5^* \right\}.$$

The next theorem determines the 5-adic group G_5.

Theorem 8.1. G_5 *is the inverse image in* $GL_2(Z_5)$ *of* $G(5)$, *in other words 5 is stable.*

Proof. Let V be the group of matrices which are

$$\equiv \begin{pmatrix} 1 & 0 \\ 0 & u \end{pmatrix} \mod 5, \qquad \text{some } u \in Z_5^* .$$

We use the refinement Lemma 2. We let $U = G_5$ so $U_1 = U \cap V_1$ consists of those elements of U which are $\equiv I \pmod 5$. Furthermore, V_1 is the full group $I + 5M_5$. It will suffice to prove:

There exist four elements a_i ($i=1,\cdots,4$) in G_5 such that

$$a_i \equiv I + 5X_i \pmod{25},$$

and X_1, X_2, X_3, X_4 are linearly independent mod 5.

We exhibit such elements by means of Frobenius elements.

From tables, for $p = 13$ we get $t_{13} = 4$, so G_5 contains a matrix σ_{13} which is a root of

$$\Phi_{13}(X) = X^2 - 4X + 13 .$$

Note that

$$\Phi_{13}(X) \equiv (X-6)(X+2) \pmod{25},$$

and $-6 \not\equiv 2 \pmod 5$. We can therefore diagonalize σ_{13} over Z_5, and we assume our coordinates so chosen that

$$\sigma_{13} = \begin{pmatrix} 6 + 25a & 0 \\ 0 & -2 + 25b \end{pmatrix} .$$

We then obtain

$$\sigma_{13}^4 \equiv I + 5 \begin{pmatrix} -1 & 0 \\ 0 & 3 \end{pmatrix} \pmod{25} .$$

In particular, σ_{13}^4 is not scalar mod 25.

Again from tables, for the prime $p = 653$ we have $t_{653} = -41$, and σ_{653} has characteristic polynomial

$$X^2 - 41X + 653 \equiv (X+11)(X-2) \pmod{25} .$$

Since $11 \not\equiv 2 \mod 5$, it follows that σ_{653} is diagonalizable over Z_5, and since

$$11^4 \equiv (-2)^4 \equiv 16 \pmod{25}$$

we obtain
$$\sigma_{653}^4 \equiv 16\,I \equiv I + 5\cdot 3\,I \pmod{25},$$

which is a scalar mod 25. Using the same basis as we picked to diagonalize σ_{13}, we see that σ_{653}^4 provides us with a diagonal element with respect to this basis mod 25. Therefore, from σ_{13}^4 and σ_{653}^4 we already obtain two linearly independent matrices in U_1/U_2, corresponding to the tangent space for the diagonal matrices.

To find a third element independent from the two preceding ones, we pick σ_{31} with $t_{31} = 7$. The characteristic polynomial is

$$X^2 - 7X + 31 \equiv (X-1)^2 \pmod{5}.$$

From the fact that $G(5)$ is a diagonal group, and $(\sigma_{31} - I)^2 \equiv 0 \pmod{5}$, we may write σ_{31} in the form

$$\sigma_{31} = I + 5Z, \qquad Z \in \text{Mat}_2(\mathbb{Z}_5),$$

and we see that Z satisfies the characteristic equation

$$Z^2 - Z + 1 = 0,$$

which is irreducible mod 5. Consequently Z is not triangular with respect to any basis. Since we already have the diagonal elements, there exists a third element of the form

$$I + 5 \begin{pmatrix} 0 & x \\ y & 0 \end{pmatrix} \pmod{25} \qquad x, y \not\equiv 0 \pmod{5}$$

linearly independent from the others mod 5.

On the other hand, the space of matrices $Y \pmod 5$ such that $I + 5Y$ belongs to $G \pmod{25}$ is invariant under conjugation by G. If we conjugate by

$$\begin{pmatrix} 1 & 0 \\ 0 & u \end{pmatrix},$$

we see that matrices

$$\begin{pmatrix} 0 & u^{-1}x \\ uy & 0 \end{pmatrix}$$

are such elements Y. From this it is immediate that we can get a fourth linearly independent matrix Y (mod 5), thereby concluding the proof.

Next we list more facts to be used about $X_0(11)$.

It is isomorphic to the Tate curve over Q_{11}. One can verify this by using the criterion for multiplicative reduction, cf. Tate [T], §6, Theorem 5 (the tangents at the double point are rational over Q_{11}). The order of its j-invariant is -5.

The equation shows that the curve has good reduction for all primes except 11. Serre [S 2] has proved that

$$G(\ell) \approx GL_2(F_\ell) \text{ for all primes } \ell \neq 5 \; .$$

All of this shows that $X_0(11)$ satisfies the hypotheses of the next theorem. Observe already that by the refinement lemma, we can conclude that

$$G_\ell = GL_2(Z_\ell), \qquad \ell > 5 \; .$$

We want to see that the image of G in $GL_2(Z_2) \times GL_2(Z_{11})$ is Serre's subgroup. In his paper, Serre asserts a similar statement for several curves (5.5.6, 5.5.7, 5.5.8, see p. 311), and he kindly communicated a proof to us. We reproduce his arguments in the following theorem, for the convenience of the reader.

Theorem 8.2. *Let* A *be an elliptic curve over the rationals. Let* L *be a set of primes containing* 2 *and an odd prime* $q \geq 5$. *Assume:*

(i) A *has good reduction at all* $\ell \in L$, $\ell \neq q$.

(ii) A *is isomorphic to the Tate curve over* Q_q, *and*

$$\text{ord}_q \, j_A$$

is not divisible by 2 *or* 3.

(iii) $G(\ell) \approx GL_2(F_\ell)$ *for all* $\ell \in L$.

Then the image of

$$\rho_L : G \longrightarrow \prod_{\ell \in L} GL_2(Z_\ell)$$

is Serre's subgroup of index 2, *namely*

$$\rho_L(G) = S_{2q} \times \prod_{\ell \neq 2, q} GL_2(Z_\ell) \; .$$

Proof. By Theorem 6.1, we know that $2 \cdot 3 \cdot q$ splits ρ, and the $GL_2(Z_\ell)$ split off for $\ell \geq 5$, $\ell \neq q$. Without loss of generality, we may therefore assume that L consists of 2, q and possibly 3. For the sake of concreteness, let us assume that $3 \in L$.

Using the isomorphism with the Tate curve over Q_{11}, one sees that G_2 (resp G_3) contains the matrix (unipotent)
$$\begin{pmatrix} 1 & 1 \\ 0 & 1 \end{pmatrix}.$$

By (iii) and the fact that the image of $GL_2(Z_\ell)$ under the determinant map is Z_ℓ^*, one sees easily that even for $\ell = 2$ or 3 one has
$$G_\ell = GL_2(Z_\ell).$$

We leave this easy part to the reader.

We now prove that the map
$$\rho_{2,3} : G \longrightarrow GL_2(Z_2) \times GL_2(Z_3)$$
is surjective. By Goursat's lemma, this amounts to determining the possible isomorphisms from a quotient of $GL_2(Z_2)$ with a quotient of $GL_2(Z_3)$. The argument concerning the unipotent matrix above can again be used to see that the image of $\rho_{2,3}$ contains
$$\begin{pmatrix} 1 & Z_2 \\ 0 & 1 \end{pmatrix} \times \{1\} \quad \text{and} \quad \{1\} \times \begin{pmatrix} 1 & Z_3 \\ 0 & 1 \end{pmatrix}.$$

Since this image is invariant by conjugation, it contains
$$SL_2(Z_2) \times \{1\} \quad \text{and} \quad \{1\} \times SL_2(Z_3).$$

Since
$$\det_{2,3} : G \longrightarrow Z_2^* \times Z_3^*$$
is surjective, it follows that the image of $\rho_{2,3}$ is the full product
$$GL_2(Z_2) \times GL_2(Z_3).$$

Finally, we determine the image of
$$\rho_{2,3,q} : G \longrightarrow [GL_2(Z_2) \times GL_2(Z_3)] \times GL_2(Z_q).$$

Since the order $q \geq 5$ does not occur in the orders of (a Jordan-Holder decomposition of) $GL_2(Z_2) \times GL_2(Z_3)$, and since $SL_2(F_q)/\pm 1$ is simple, it follows that in Goursat's lemma, any possible quotient of $GL_2(Z_q)$ must have order prime to q and must be abelian. One sees easily that:

$$GL_2(Z_\ell)^{ab} = Z_\ell^* \quad \text{(via det)} \quad \text{if } \ell \geq 3$$

$$GL_2(Z_2)^{ab} = \{\pm 1\} \times Z_2^* \quad \text{(via } \varepsilon \times \text{det)} \quad \text{if } \ell = 2,$$

where ε is the homomorphism

$$\varepsilon : GL_2(F_2) = S_3 \longrightarrow \{\pm 1\}.$$

Let

$$\psi : G_2 \times G_3 \times G_q \longrightarrow G_2^{ab} \times G_3^{ab} \times G_q^{ab}$$

be the factor commutator group mapping. As we have seen, the group G must contain $(G_2 \times G_3)'$ and G_q'. Thus $\psi(G)$ is a subgroup

$$\psi(G) \subset \{\pm 1\} \times Z_2^* \times Z_3^* \times Z_q^*,$$

and because the roots of unity are contained in the field of division points, we know that $\psi(G)$ must project onto

$$Z_2^* \times Z_3^* \times Z_q^*.$$

On the other hand, $\psi(G)$ is necessarily contained in the subgroup consisting of elements

$$(\pm 1, u_2, u_3, u_q)$$

such that

$$\left(\frac{u_q}{q}\right) = \pm 1,$$

which has index 2. Therefore $\psi(G)$ is equal to this subgroup. This implies that

$$\rho_L(G) = S_{2q} \times GL_2(Z_3),$$

thereby proving the theorem.

In the application, we let

$$L = \{2, 3, \ell \geq 7\}, \qquad q = 11.$$

Theorem 6.2 gives us the structure of G_L for the Shimura curve. We then have to see how it mixes with G_5, which amounts to determining the Goursat subgroups. We shall need additional notation for this, and the result is stated at the end, as Theorem 8.3.

We have

$$G \subset G_5 \times G_L, \qquad \text{and} \qquad G_L = S_{2q} \times \prod_{\substack{\ell \neq 2, q \\ \ell \in L}} GL_2(Z_\ell).$$

We let H_5 and H_L be the corresponding Goursat subgroups, giving an isomorphism

$$G_5/H_5 \approx G_L/H_L,$$

i.e. giving the identification of G_5 and G_L on the common subfield $K_5 \cap K_L$, which is a finite extension of Q. We shall determine explicitly what this subfield is, and what the above isomorphism is.

We have an exact sequence

$$1 \longrightarrow U_5 \longrightarrow G_5 \longrightarrow Z(4) \longrightarrow 1,$$

where $U_5 = 1 + 5M_5$. The map $H_5 \to Z(4)$ is surjective because $Q(\mu_5)$ is disjoint from K_L (the field of fifth roots of unity $Q(\mu_5)$ is ramified only at 5, and K_L is unramified at 5). Thus we conclude:

$(G_5 : H_5)$ *is a power of* 5.

Lemma 1. G_5/H_5 *is abelian, and* H_L *contains*:
(i) $SL_2(Z_\ell)$ *for* $\ell \neq 2, 5, 11 = q$.
(ii) $(1 + 2M_2) \times (1 + qM_q)$ *as subgroup of* $S_{2q} \subset H_L$.
Furthermore, $S_{2q}(2q)$ *contains the product*

$$E(2) \times SL_2(q),$$

where $E(2)$ *is the group of even elements at* 2.

Proof. The first two parts (i) and (ii) follow from the standard Jordan-Holder consideration, either because of the simplicity of $SL_2(F_\ell)/\pm 1$ for $\ell \geq 5$, or because primes dividing the orders of certain subgroups on the right do not occur in H_5, that is, are not equal to 5. The last statement follows similarly, noting that the even elements form a subgroup of order $3 \neq 5$, and that $SL_2(q)$ has a Jordan-Holder decomposition of a simple group and a group of order $2 \neq 5$. The prime 5 occurs only once in the part of G_L^{ab} ramified only at 11. It follows at once that G_5/H_5 has order 1 or 5. This proves the lemma.

Lemma 2. We have
$$H_L = H_{2q} \times \prod_{\ell \neq 2,5,11} GL_2(Z_\ell)$$
and $(S_{2q} : H_{2q}) = 1$ or 5.

Proof. The group G_5/H_5 is the Galois group of a cyclic extension of the rationals which is ramified only at 11 or possibly 5, and of degree 1 or 5. On the other hand, G_L/H_L is the Galois group of an abelian extension which is unramified at 5. Hence the common subfield $K_5 \cap K_L$ must be the 5-part of the field of 11-th roots of unity,
$$Q(\mu_{11}),$$
or it must be trivial. This proves the lemma, while also providing additional information.

We must determine $K_5 \cap K_{22}$, or in other words $H_{2q} = H_{2 \cdot 11}$. For this we need to make some mappings more explicit.

We know that G_5 is the group of matrices
$$a = \begin{pmatrix} 1 + 5a & 5b \\ 5c & u \end{pmatrix}$$
where $a, b, c \in Z_5$ and $u \in Z_5^*$. There is a homomorphism
$$\psi : a \longmapsto a(a) = a \pmod{5}$$
whose kernel contains the commutator subgroup G_5', and consists of the matrices
$$\begin{pmatrix} 1 + 25a & 5b \\ 5c & u \end{pmatrix}.$$

The homomorphism ψ cannot factor through

$$\det_5 : G_5 \longrightarrow Z_5^*,$$

because, for instance, we have $\psi(\sigma) \neq 0$ and $\det \sigma = 1$ if σ is the matrix

$$\begin{pmatrix} 6 & 0 \\ 0 & 1/6 \end{pmatrix}.$$

The map $\psi \circ \mathrm{pr} : G_{2 \cdot 11} \times G_5 \to Z(5)$ which factors through the projection on G_5 is abelian. Let $M = 2 \cdot 11 \cdot 5$. Let

$$\rho_M : G \longrightarrow GL_2(Z_2) \times GL_2(Z_{11}) \times GL_2(Z_5)$$

be the partial representation on those three factors. Then $\psi \circ \mathrm{pr} \circ \rho_M$ is abelian on G, and hence factors through G'_M, which is the intersection of G with SL_2. This means that $\psi \circ \mathrm{pr} \circ \rho_M$ factors through the determinant map on G. In other words, the following diagram commutes on G, with some homomorphism $\lambda_{22,5}$.

$$(1 \times H_5) \subset G \xrightarrow{\rho_M} G_{22} \times G_5 \xrightarrow{\psi \circ \mathrm{pr}} Z(5)$$

$$\det \downarrow \qquad \nearrow \lambda_{22,5}$$

$$Z_{22}^* \times Z_5^*$$

In symbols, this means

$$\psi \circ \mathrm{pr} \circ \rho_M = \lambda \circ \det \circ \rho_M.$$

We apply this to the subgroup $\{1\} \times H_5$ which is contained in G, and conclude:

Lemma 3. *We have $H_5 \neq G_5$ and $H_5 \subset \mathrm{Ker}(\psi - \lambda \circ \det)$, where $\lambda = \lambda_5$ is the restriction of $\lambda_{22,5}$ to the second factor. The map*

$$\psi - \lambda \circ \det : G_5/H_5 \longrightarrow Z(5)$$

is an isomorphism.

Since, as we have already remarked, G_5/H_5 is abelian of order 5, it now follows that G_L/H_L is abelian of order 5, and factors through the projection on

G_{11} and the determinant map to give a natural isomorphism

$$\phi_{11} \circ \mathrm{pr}_{11} : G_L/H_L \approx Z(11)^*/\pm 1 ,$$

as Galois group of the 5-part of the cyclotomic field $Q(\mu_{11})$. Thus our task is to find explicitly the isomorphism

$$\phi_5 : G_5/H_5 \longrightarrow Z(11)^*/\pm 1 .$$

The group on the right is cyclic, and we can select ± 2 as a generator (primitive root). We can now state the main theorem tying up all the lemmas together.

Theorem 8.3. *The homomorphism λ is trivial. The isomorphism* $G_5/H_5 \to Z(11)^*/\pm 1$ *is given by*

$$\phi_5 : \sigma \longmapsto (\pm 2)^{\psi(\sigma)} .$$

The number $M = 2 \cdot 25 \cdot 11$ *splits and stabilizes* ρ. *The group* G_{110} *consists of the elements* $(\sigma_5, \sigma_2, \sigma_{11})$ *such that*

$$\phi_5(\sigma_5) = \phi_{11}(\sigma_{11}) \qquad \text{and} \qquad (\sigma_2, \sigma_{11}) \in S_{2 \cdot 11} .$$

Proof. We have seen that the map

$$\psi_1 = \psi - \lambda \circ \det : G_5/H_5 \longrightarrow Z(5)$$

is an isomorphism, and identifying $Z(5)$ with its action on the 11-th roots of unity, we may write

$$\psi_1(\sigma) = \det \rho_{11}(\sigma)/\pm 1 .$$

Note that λ is defined on $Z_5^*(25)$, because $\mathrm{Im}\,\lambda$ is cyclic and the group $1 + 5Z_5$ is cyclic. There is a unique element $w \in Z(5)$ such that

$$\lambda \circ \det(\sigma) = w \frac{(\det \sigma)^4 - 1}{5} .$$

We shall prove that $w = 0$. Let

$$\sigma = \begin{pmatrix} 1 + 5a & 5b \\ 5c & u \end{pmatrix} , \qquad \psi(\sigma) = a .$$

Then $\det \sigma = u(1+5a) \pmod{25}$, $\operatorname{tr} \sigma = 1+u+5a$, and

$$(1-u)a = \frac{\operatorname{tr} \sigma - \det \sigma - 1}{5} \pmod 5.$$

If $\det \sigma \not\equiv 1 \pmod 5$ then

$$\psi(\sigma) = a = \frac{\operatorname{tr} \sigma - \det \sigma - 1}{5(1-u)} = \frac{\operatorname{tr} \sigma - \det \sigma - 1}{5(1-\det \sigma)} \pmod 5.$$

Thus for $p \not\equiv 1 \pmod 5$ we can calculate $\psi(\sigma_p)$ from $t_p = \operatorname{tr} \sigma_p$ and $p = \det \sigma_p$, and in particular obtain the following table.

$p = \det \sigma_p$	t_p	$\psi_1(\sigma_p) = \det \sigma_p$ in $Z(11)^*/\pm 1$	$\psi(\sigma_p)$	$\dfrac{(\det \sigma_p)^4 - 1}{5}$
2	-2	± 2	1	-2
13	4	± 2	1	2

The table shows that $\psi_1(\sigma_2) = \psi_1(\sigma_{13})$ and $\psi(\sigma_2) = \psi(\sigma_{13})$, and therefore $\lambda \circ \det (\sigma_2) = \lambda \circ \det (\sigma_{13})$. Since the entries in the last column of the table are not congruent (mod 5), w must be 0. Since $\psi_1 = \pm 2$ when $\psi = 1$, the isomorphism ϕ_5 must be as given in the statement of the theorem.

Having determined the Galois group, we saw that

$$M = 2 \cdot 11 \cdot 5^2$$

stabilizes and splits ρ. We are then in a position to determine the constant at the bad primes.

Theorem 8.4. Let $q = 11$. For the curve $X_0(11)$, we have

$$F_M(t) = F_{2q}(t) F_{25}(t).$$

Proof. It suffices to prove that if t_1 is the residue class of t mod $M_1 = 2 \cdot 11$ and t_2 is the residue class of t mod $M_2 = 25$, then

$$|G(22)_{t_1}| \, |G(25)_{t_2}| = 5 \, |G(22 \cdot 25)_t| \, .$$

Let
$$\phi_1 : G(22) \longrightarrow Z(5) \quad \text{and} \quad \phi_2 : G(25) \longrightarrow Z(5)$$
be the two homomorphisms such that $G(M) = \text{Ker } \lambda$, where
$$\lambda = \phi_1 \otimes 1 - 1 \otimes \phi_2 \, .$$

We have a correspondence
$$G(M)_t \longrightarrow G(22)_{t_1} \times G(25)_{t_2}$$
which associates
$$\sigma \longmapsto \left(\sigma_1, \sigma_2 + \begin{pmatrix} 5a & 0 \\ 0 & -5a \end{pmatrix} \right),$$
with $a \in Z(5)$. It is clear that this association establishes a one-to-five correspondence which makes the theorem obvious.

Remark. The value $F_{2q}(t)$ is exactly the same as that found in §5 for the Serre curves, in the tables.

There remains to determine $F_{25}(t)$.

Theorem 8.5. *We have:*
$$F_{25}(t) = \begin{cases} 5/4 & \text{if } t \not\equiv 1 \bmod 5 \\ 0 & \text{if } t \equiv 1 \bmod 5 \, . \end{cases}$$

Proof. The value 0 is clear. For the others, note that we have for $t, s \not\equiv 1 \pmod 5$:
$$|G(25)_t| = |G(25)_s| \, ,$$
arising from the bijection
$$\sigma \longmapsto \sigma + \begin{pmatrix} 0 & 0 \\ 0 & s-t \end{pmatrix} \, .$$

The desired values then follow at once from the definitions.

PART II

IMAGINARY QUADRATIC DISTRIBUTION

We let k be an imaginary quadratic field, with discriminant D, so $k = Q(\sqrt{D})$. We let w be the number of roots of unity in k, and we let h be the class number. We let o be the integers of k.

Let A be an elliptic curve over the rationals. For each prime p where A has good reduction, we have a Frobenius endomorphism π_p, and we want to describe conjecturally a probabilistic model for which the sequence of traces t_p such that $Q(\pi_p) = k$ forms a random sequence. We work entirely in the Galois theory setting of division points, and thus axiomatize the situation. The matter is briefly reviewed in §3. We shall see in §5 that the probability that $Q(\pi_p) = k$ is conjecturally asymptotic to

$$C(k, A) \frac{1}{2\sqrt{p}},$$

for some constant $C(k, A) > 0$. The conjecture implies that the number of primes $p \leq x$ for which $Q(\pi_p) = k$ is asymptotic to

$$C(k, A) \pi_{\frac{1}{2}}(x),$$

where

$$\pi_{\frac{1}{2}}(x) = \sum_{p \leq x} \frac{1}{2\sqrt{p}} \sim \frac{\sqrt{x}}{\log x}.$$

We let K be the field of division points. We go through similar steps as for the fixed trace distribution discussed in Part I, but in a more complicated setting. The complications arise from at least two factors:

(a) The presence of units in the imaginary quadratic field k, which always cause ambiguity in the set of generators of an ideal.

(b) The possible dependence of the GL_2-extension K with the maximal abelian extension k_{ab} of k. Usually these two fields intersect in the field generated over the rationals by all roots of unity, i.e. Q_{ab} (see Theorem 3.1), but it may happen that the intersection is bigger, by a finite extension which is an important invariant of the situation, and must be taken into account. So must the intersection of the GL_2-extension and the Hilbert class field H of k.

It is therefore natural to work with the composite extension Kk_{ab}, discussed in §3, giving rise to the probabilistic correspondence.

Practically, for the probabilistic model the above factor means that instead of parametrizing the probabilistic fiber at each prime by a single integer, we must now use pairs of integers. The first is a random variable for the trace of Frobenius in the GL_2-extension, and the second is a random variable for the trace of Frobenius in k. These will not be independent!

One can also consider the case of fixed trace from the quadratic field. For elliptic curves having complex multiplication, Mazur's "anomalous" primes (those with fixed trace $t_p = 1$, see [Ma]) happen to lie in certain quadratic progressions, for which Hardy and Littlewood had conjectured the asymptotic behavior. Both for later use, and also because it shows in a simple case how the arithmetic of the quadratic field affects the distribution, we recover the Hardy-Littlewood conjecture independently by making up a probabilistic model similar to the other cases. As an example, when $k = Q(i)$, the elements of trace 2 are those of the form $1 + ni$, and this special case gives the conjectured asymptotic behavior of those primes p which are of the form $n^2 + 1$. There is hardly any need to remind the reader that it is still unknown if in fact there exist infinitely many such primes.

Our main problem is to describe the constant $C(k, A)$.

We first show how one can define the constant $C(k, A)$ by taking a limit from finite levels, in a manner compatible with the Sato-Tate, Hecke, and Tchebotarev density properties. This is done in §4 and §5. We then show how the constant can be written as a quotient of integrals.

The denominator involves essentially the index of the Galois group $Gal(Kk_{ab}/H)$ in the "generic" Galois group, and the measure of this "generic" Galois group. It is convenient for our purposes to take additive Haar measures, so the measures of the multiplicative groups have to be computed.

The numerator involves the direct image of Haar measure into the space of conjugacy classes of integral Cartan elements corresponding to k. In §7, §8, §9 the measures and density functions involved are determined explicitly, and tabulated. Langlands suggested to us that what we were doing could be interpreted as computation of Harish transforms, cf. the comments at the end of §6. Theorem 7.1, Theorem 8.1, and the lemmas following Theorem 8.1 give a complete and systematic way of evaluating Harish transforms, and their integrals. The Harish transform in a neighborhood of the identity is computed in Sally-Shalika [S-Sh], but the context here is sufficiently different and the need for more systematic tables such that it was pointless to refer further to the literature.

Both the numerator and denominator admit an infinite product decomposition into local ℓ-factors. These are rational functions in $r = r(\ell) = 1/\ell$. If one writes the global constant $C(k, A)$ in terms of the discriminant, then these factors are of the form

$$1 + O(r^2),$$

and their product is absolutely convergent. If one writes the constant $C(k, A)$ in terms of the class number, then the ℓ-th factor is of the form

$$\left(1 - \left(\frac{k}{\ell}\right)\frac{1}{\ell}\right)^{-1} (1 + O(r^2))$$

and their product gives the value $L(1, \chi)$, times an absolutely convergent product. With our arguments, it is this second form which comes naturally. The constant is inversely proportional to $\sqrt{|D|}$.

In §10, §11 we consider the same special cases as in Part I, i.e. elliptic curves of Serre type, and $X_0(11)$, for which we obtain numerical values for the constant. This allows us to compare the predicted values with actual values for the asymptotic behavior of the primes in question, computed by machine. The fit is quite good on the whole, and is discussed in Part IV.

PART II

IMAGINARY QUADRATIC DISTRIBUTION

THE FIXED TRACE CASE

1.	Fixed traces from the quadratic field	77
2.	Computation of the constant for fixed trace	84

THE MODEL FOR THE MIXED CASE

3.	The mixed Galois representations	91
4.	The probabilistic model	104
5.	The asymptotic behavior	108
6.	The finite part of the constant as a quotient of integrals	112

COMPUTATIONS OF HARISH TRANSFORMS

7.	Haar measure under the trace-determinant map on Mat_2. General formalism.	123
8.	Relations with the trace-norm map on k	133
9.	Computation of C_ℓ for almost all ℓ	141
10.	The constant for Serre curves, $K \cap k_{ab} = Q_{ab}$	143
11.	The constant for $X_0(11)$	149

THE FIXED TRACE CASE

§1. Fixed traces from the quadratic field

In the probabilistic model, there is a random variable for each prime p, which will range over the integers. The probability function will again have a factor at infinity, and a factor at the finite primes.

We first describe the density function at infinity, which in the present instance is none other than the function of Hecke giving equidistribution of primes in sectors. We let

$$g''(\xi) = w \frac{1}{\pi} \frac{1}{\sqrt{1-\xi^2}}.$$

Then g'' is the distribution function of primes in k at infinity. The trace from k to Q will be abbreviated sometimes without subscript, e.g.

$$\text{Tr} = \text{Tr}_{k/Q}.$$

Because g'' blows up somewhat at the end points of the interval, it is convenient to define a truncation. We redefine $\xi(t)$ when $|t| < 2\sqrt{p}$ by letting

$$\xi(t) = \min\left\{\frac{t}{2\sqrt{p}},\ 1 - \frac{1}{2\sqrt{p}}\right\} \qquad \text{if } t > 0$$

$$\xi(t) = \max\left\{\frac{t}{2\sqrt{p}},\ -1 + \frac{1}{2\sqrt{p}}\right\} \qquad \text{if } t < 0.$$

We let P_k be the set of primes p which split completely in k,

$$p\mathfrak{o} = \mathfrak{p}\bar{\mathfrak{p}},$$

and such that the prime ideals $\mathfrak{p}, \bar{\mathfrak{p}}$ are principal, say $\mathfrak{p} = (\pi)$. The generator π is determined up to a unit in $\mathfrak{o}$.

Remark. By Tchebotarev, the density of primes which split completely in k is 1/2. Therefore the density of primes which have the above property is equal to 1/2h.

We let $P_k(x)$ be the set of primes $p \in P_k$ with $p \leq x$. For any interval J contained in $[-1, 1]$, Hecke's theorem implies the density property:

$$\int_J g''(\xi) d\xi = \lim_{x \to \infty} \frac{1}{|P_k(x)|} \sum_{p \in P_k(x)} \#\{\pi \in \mathfrak{o}, \mathrm{Tr}\,\pi \in J, \pi\bar\pi = p\}.$$

In other words, $\xi(\mathrm{Tr}\,\pi)$ for prime elements π as above is distributed according to the function g'' on the interval $[-1, 1]$. Note that such primes are counted without making any identification, e.g. $g(\xi(\mathrm{Tr}\,\pi)) = g(\xi(\mathrm{Tr}(-\pi)))$, but π and $-\pi$ are counted as distinct.

To deal with the presence of roots of unity in k, and to fix representatives of elements modulo such units, we can restrict our attention to smaller intervals. We say that a subinterval J of $[-1, 1]$ is restricted if it is the projection of an arc in the upper half of the unit circle having length π/w. We let g''_J be the function which is equal to g'' on the interval J, and equal to 0 outside that interval.

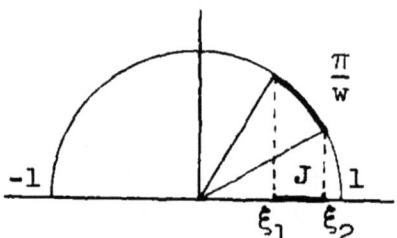

When J is restricted as above to be the projection of an arc with angle π/w, then we can also write the density relation in the form

$$\int_J g''(\xi) d\xi = \lim_{x \to \infty} \frac{\#\{p \in P_k(x), \mathrm{Tr}\,\pi \in J \text{ for some } \pi \text{ with } \pi\bar\pi = p\}}{|P_k(x)|}$$

We now pass to congruence conditions.

For a positive integer M we let $\mathfrak{o}(M) = \mathfrak{o}/M\mathfrak{o}$, and we let $\mathfrak{o}^*(M)$ be the image of $\mathfrak{o}^*$ in $\mathfrak{o}/M\mathfrak{o}$, actually in $\mathfrak{o}(M)^*$. If $M \geq 3$ then $\mathfrak{o}^* \to \mathfrak{o}^*(M)$ is injective, and consequently we also identify $\mathfrak{o}^*$ with the image $\mathfrak{o}^*(M)$. The factor group

$$o(M)^*/o^*$$

is going to play an important role in the sequel.

We let $P_{k,M}$ be the subset of primes of P_k which do not divide M, and $P_{k,M}(x)$ the subset of those primes $\leq x$. Let $\mathfrak{p}$ be a prime ideal above p. Given an integer s, we define

$$r_M(p,s) = \#\{\pi \text{ generating } \mathfrak{p} \text{ such that } \text{Tr}\,\pi \equiv s \bmod M\}.$$

We fix a positive integer $M \geq 3$. We define the function $F_M(s)$ by the formula

$$\frac{w}{M} F_M(s) = \lim_{x \to \infty} \frac{1}{|P_{k,M}(x)|} \sum_{p \in P_{k,M}(x)} r_M(p,s).$$

Except for taking into account the multiplicity of elements in a given class mod roots of unity o^*, the function F_M essentially counts the density relative to P_k of those primes p such that $\text{Tr}(\pi_p)$ mod M is a given congruence class mod M. In §2 we shall give a simple expression for $F_M(s)$ by using the Tchebotarev density theorem. For the moment, we pursue the probabilistic considerations.

From the definitions we immediately obtain

(1) $$\sum_{s \bmod M} \frac{1}{M} F_M(s) = 1.$$

We shall see below that

$$\lim_M F_M(s) = F(s)$$

exists.

We have fixed the interval J (corresponding to a sector of width π/w), and M. For each prime p we let the fiber of the probabilistic model at p be Z. We let the measure $\mu_{p,J}$ be represented with respect to counting measure by a function

$$f_{M,J}(s,p) \geq 0,$$

which we assume of the form

PR 1. $$f_{M,J}(s,p) = c_p g''_J(\xi(s,p)) F_M(s),$$

where c_p is a constant such that

(2) $$\sum_s f_{M,J}(s,p) = 1.$$

By considering the value

$$1 = \int_{-1}^{1} g''_J(\xi'') d\xi'',$$

and approximating Riemann sums

$$\frac{1}{2\sqrt{p}} \sum_s g''_J(\xi(s,p)), \qquad \text{for } p \to \infty,$$

it follows trivially, as in the first part, that we have the asymptotic relation

$$c_p \sim \frac{1}{2\sqrt{p}}.$$

For a fixed congruence class we have

$$\Pr\{y_p \equiv s_0 \bmod M\} = f_M(s_0, p) = c_p g'(\xi(s_0, p)) F_M(s_0).$$

Since

$$\xi(s_0, p) = s_0 / 2\sqrt{p},$$

it follows that

$$\lim_p g''(\xi(s_0, p)) = g''(0) = w/\pi.$$

Originally we had taken the density of p relative to P_k. We want the density relative to all primes, so we define the constant

$$\boxed{C(s_0, k) = \frac{1}{2h} \frac{w}{\pi} F(s_0)}$$

where h is the class number. The conjecture is that the number of primes in $P_k(x)$ with a fixed trace

$$\text{Tr}(\pi_p) = s_0$$

is given by the asymptotic formula

$$\boxed{N_{s_0, k}(x) \sim C(s_0, k) \pi_{\frac{1}{2}}(x),}$$

where

$$\pi_{\frac{1}{2}}(x) = \sum_{p \leq x} \frac{1}{2\sqrt{p}} \sim \frac{\sqrt{x}}{\log x}.$$

Example 1. Take $s_0 = 2$ and $k = Q(i)$. Then the elements with trace 2 in o are of the form

$$1 + ni$$

with some integer n, and therefore $N_{2,Q(i)}(x)$ is the number of primes $p \leq x$ such that p is of the form

$$p = n^2 + 1.$$

We shall determine the value of $F(2)$, thus recovering the Hardy-Littlewood conjecture concerning the asymptotics of such primes. From the tables of the next section, we find that the constant in this case is

$$C(2, Q(i)) = \frac{4}{\pi} \prod_{(k/\ell)=1} \frac{\ell(\ell-2)}{(\ell-1)^2} \prod_{(k/\ell)=-1} \frac{\ell^2}{\ell^2-1}.$$

On the other hand, let $L(1, \chi)$ be the value of the L-series of k with the non-trivial character. Then

$$L(1, \chi) = \frac{\pi}{4} = \prod_{(k/\ell)=1} \frac{1}{1-1/\ell} \prod_{(k/\ell)=-1} \frac{1}{1+1/\ell}.$$

(The products are written formally by abuse of notation. They should be combined and ordered by increasing ℓ, otherwise they don't converge.) From this we find

$$C(2, Q(i)) = \prod_{\ell \text{ odd}} \left(1 - \left(\frac{-1}{\ell}\right)\frac{1}{\ell-1}\right)$$

which is the constant of Hardy-Littlewood.

The 5,000th prime is 48,6811, and direct computation gives $\pi_{\frac{1}{2}}(48,611) \approx 26.4$. Putting this in our asymptotic formula (and taking enough terms in the product for $C(2, Q(i))$ to make the relative error less than .01) we get 36.2 as an estimate for $N_{2,Q(i)}(48,611)$. It is trivial with the computer (and not hard by hand) to count the numbers of the form $n^2 + 1 \leq 48,611$ that are prime, and there turn out to be 37 of them. Similar calculations and counts give the following table.

$N_{s,k}(48,611)$, $k = Q(i)$

s	estimated	actual		s	estimated	actual
2	36.2	37		6	24.2	28
4	36.2	38		12	24.2	26
8	36.2	33				
16	36.2	33				
32	36.2	38				

There is good agreement between estimates and actual counts.

Example 2. There are prime elements of trace 1 in the quadratic field with discriminant D if and only if $D = 1 - 4m$, with m odd. It is then easy to check that the rational prime p has a complex factor with trace 1 if and only if it lies in the quadratic progression

$$m - Dx - Dx^2 .$$

For $m = 1, 3, 5, 9, 11$ one gets $D = -3, -11, -19, -35, -43$. The class number h is 2 for $D = -35$, and 1 for the others. The number of units w is 6 for $D = -3$, and 2 for the others. Estimated and actual counts are given in the table below.

$N_{1,k}(48,611)$, $k = Q(\sqrt{D})$

D	estimated	actual
−3	51.3	47
−11	8.9	10
−19	12.1	13
−35	6.0	7
−43	13.5	13

Agreement is again quite good. Since the occurrence of trace 1 has special interest [Ma], we include a table of the primes occurring in these five cases.

-3	-11	-19	-35	-43
7	3	5	79	11
19	223	43	709	97
37	619	233	6379	269
61	1213	1069	13309	1301
127	5569	1373	28429	3881
271	13093	2969	34729	6719
331	18043	3463	41659	9041
397	26953	8783		10331
547	30319	9619		14717
631	45763	14369		23747
919		21323		27961
1657		28163		30197
1801		34319		42667
1951				
2269				
2437				
2791				
3169				
3571				
4219				
4447				
5167				
5419				
6211				
7057				
7351				
8269				
9241				
10267				
11719				
12097				
13267				
13669				
16651				
19441				
19927				
22447				
23497				
24571				
25117				
26227				
27361				
33391				
35317				
42841				
45757				
47251				

Primes ≤ 48,611 with a principal factor of trace 1 in $Q(\sqrt{D})$, for $D = -3, -11, -19, -35, -43$.

§2. Computation of the constant for fixed trace

We use class field theory over k, and the notation established here concerning the maximal abelian extension of k will also be used later when we mix the situation with the GL_2 extension.

We let k_{ab} be the maximal abelian extension of k. Its Galois group is isomorphic to the finite part of the group of idele classes of k. We shall be interested in the Galois group $\mathfrak{A}$ of k_{ab} over the Hilbert class field H. Let U denote the unit ideles of k, from which we project out the component at infinity. Thus

$$U = \prod_\ell \mathfrak{o}_\ell^*, \qquad U_\ell = \mathfrak{o}_\ell^*, \qquad \text{and} \quad \overline{U} = U/\mathfrak{o}^*.$$

Then we have the class field isomorphism

$$\mathfrak{A} = \text{Gal}(k_{ab}/H) \approx Uk^*/k^* \approx (U/(U \cap k^*) = U/\mathfrak{o}^*,$$

where $\mathfrak{o}^*$ is the group of units in $\mathfrak{o}$, i.e. the roots of unity. In this manner we get a representation

$$\rho'' : \text{Gal}(k_{ab}/H) \longrightarrow \left(\prod_\ell \mathfrak{o}_\ell^*\right)/\mathfrak{o}^* = \overline{U}.$$

We let $\mathfrak{o}(M) = \mathfrak{o}/M\mathfrak{o}$, and we let $\mathfrak{o}^*(M)$ be the image of $\mathfrak{o}^*$ in $\mathfrak{o}/M\mathfrak{o}$, actually in $\mathfrak{o}(M)^*$. We have a natural homomorphism

$$\overline{U} = U/\mathfrak{o}^* \longrightarrow \mathfrak{o}(M)^*/\mathfrak{o}^*(M).$$

If $M \geq 3$, then the map $\mathfrak{o}^* \to \mathfrak{o}^*(M)$ is injective (no two roots of unity are congruent mod M). We let

$$\rho''_{(M)} : \text{Gal}(k_{ab}/H) = \mathfrak{A} \longrightarrow \mathfrak{o}(M)^*/\mathfrak{o}^*(M) = \overline{U}(M).$$

be the composition of ρ'' with the canonical homomorphism mod M. This reduction mod M of the representation has a kernel, whose fixed field is denoted by $k_{ab}(M)$, and we let $\mathfrak{A}(M)$ denote its Galois group over H, so that we obtain an isomorphism

(for $M \geq 3$),
$$\rho''_{(M)} : \mathfrak{A}(M) \longrightarrow \mathfrak{o}(M)^*/\mathfrak{o}^* .$$

As usual, the prime ideals of k not dividing M (say above $\mathfrak{p} \in P_k$) are embedded in the idele classes, whence in $U/\mathfrak{o}^*$ by mapping a prime element $\pi_{\mathfrak{p}}$ on the idele having $\pi_{\mathfrak{p}}^{-1}$ at the $\mathfrak{p}$-component, and 1 at all other components. Then
$$\rho''_{(M)}(\sigma_{\mathfrak{p}}) \equiv \pi_{\mathfrak{p}} \pmod{M} ,$$
because the Artin symbol of that idele has the same effect as the idele having $\pi_{\mathfrak{p}}$ at all components except for component 1 at $\mathfrak{p}$.

If X is a subset of $\mathfrak{o}(M)$, and s an integer, we let X_s be the subset of X consisting of those elements u such that
$$\text{Tr } u \equiv s \pmod{M} .$$

Theorem 2.1. *If $M \geq 3$, then*
$$\frac{1}{M} F_M(s) = \frac{|\mathfrak{o}(M)^*_s|}{|\mathfrak{o}(M)^*|} .$$

Proof. Let G be the Galois group of $k_{ab}(M)$ over Q. We define a function $\lambda_M(\sigma, s)$ for $\sigma \in G$ to be 0 unless σ lies in $\mathfrak{A}(M)$, that is σ leaves H fixed, and in that case,

$\lambda_M(\sigma, s) = $ number of elements in $\rho''_{(M)}(\sigma)$ (which is a coset of $\mathfrak{o}^*$) having a trace $\equiv s \bmod M$.

Then λ_M is a function of conjugacy classes in G. We may view our previous function $r_M(p,s)$ as defined for any prime p, but equal to 0 except when p splits completely in k, and its factors are principal in k, and p does not divide M, in which case it has the value assigned to it previously. We let P be the set of all primes, and $P(x)$ the subset of those primes $\leq x$. Then

$$\frac{W}{M} F_M(s) = \lim_{x \to \infty} \frac{1}{|P_k(x)|} \sum_{p \in P_k(x)} \tau_M(p,s)$$

$$= \lim_{x \to \infty} \frac{2h}{|P(x)|} \sum_{p \in P(x)} \lambda_M(\sigma_p, s)$$

$$= \frac{2h}{|G|} \sum_{\sigma \in G} \lambda_M(\sigma, s) \qquad \text{(by Tchebotarev)}$$

$$= \frac{1}{|\mathcal{C}(M)|} |o(M)_s^*|$$

$$= \frac{W}{|o(M)^*|} |o(M)_s^*| \; .$$

This proves the theorem.

From Theorem 2.1 we can compute F_M by elementary congruence arguments, carried out in the next lemmas. The first one gives the multiplicativity over relatively prime factors. The second shows that the value of F_{ℓ^n} stabilizes at level 1 for odd primes, and at level 2 if $\ell = 2$. The third lemma gives an invariance under multiplicative translations, which reduces computations to just a few cases: 0 or 1 for odd ℓ and 0, 1, 2, 3 for even ℓ.

Lemma 1. *Suppose that* $M = M_1 M_2$ *is a product of two relatively prime factors. Then*

$$F_M(s) = F_{M_1}(s) F_{M_2}(s) \; .$$

Proof. Obvious.

Lemma 2. *Let* ℓ *be prime,* $n \geq 1$ *if* ℓ *is odd and* $n \geq 2$ *if* $\ell = 2$. *Let* s, t *be integers such that*

$$s \equiv t \bmod \ell \qquad \text{or} \qquad s \equiv t \bmod 4$$

according as ℓ *is odd or even. Then*

$$|o(\ell^n)_s^*| = |o(\ell^n)_t^*|$$

and

$$F_{\ell^n}(s) = F_{\ell^n}(t) = F_\ell(s) \qquad \text{if } \ell \text{ is odd}$$

$$F_{2^n}(s) = F_{2^n}(t) = F_4(s) \qquad \text{if } \ell = 2 \; .$$

Proof. The map
$$u \mapsto u + \frac{t-s}{2}$$
gives a bijection of $\mathfrak{o}(\ell^n)_s^*$ with $\mathfrak{o}(\ell^n)_t^*$ so the first assertion is obvious. The second follows from the first, and the fact that under the reduction map

$$\mathfrak{o}(\ell^m)^* \longrightarrow \mathfrak{o}(\ell^n)^*$$

there are $\ell^{2(m-n)}$ elements in each fiber, while for the reduction

$$Z(\ell^m) \longrightarrow Z(\ell^n)$$

there are ℓ^{m-n} elements in each fiber.

Lemma 3. *If $a \in Z(M)^*$ then $F_M(as) = F_M(s)$.*

Proof. The map $u \mapsto au$ is a bijection of $\mathfrak{o}(M)_s^*$ onto $\mathfrak{o}(M)_{as}^*$.

Putting these lemmas together, we obtain:

Theorem 2.2. *The constant $C(s_0, k)$ is given by*

$$C(s_0, k) = \frac{1}{2h} \frac{w}{\pi} F(s_0),$$

where

$$F(s_0) = F_4(s_0) \prod_{\ell \neq 2} F_\ell(s_0).$$

We must then tabulate $F_4(s_0)$ and $F_\ell(s_0)$.

Theorem 2.3. *Let ℓ be odd. The values of F_ℓ are given by the following table, where $r = r(\ell) = 1/\ell$.*

	$F_\ell(0)$	$F_\ell(1)$
$\left(\frac{k}{\ell}\right) = 1$	$\frac{1}{1-r}$	$\frac{1-2r}{(1-r)^2}$
$\left(\frac{k}{\ell}\right) = -1$	$\frac{1}{1+r}$	$\frac{1}{1-r^2}$
$\left(\frac{k}{\ell}\right) = 0$	$\frac{1}{1-r}$	$\frac{1}{1-r}$

Proof. In the first case, $o(\ell)^* \approx F_\ell^* \times F_\ell^*$. In the second case,

$$o(\ell)^* \approx F_{\ell^2}^*.$$

In the third case, if λ is a prime element above ℓ, then representatives for $o(\ell)$ are given by $a + b\lambda$ with $a, b \in Z/\ell Z$. It is then a routine matter in each case to determine the cardinality of $o(\ell)^*$, and then $F_\ell(0)$ and $F_\ell(1)$.

The case when $\ell = 2$ is slightly more complicated, especially when 2 ramifies in k, and two subcases arise. We say that the ramification is of first kind if there exists $\lambda \in o$ such that $\text{Tr}(\lambda) \equiv 2 \pmod{4}$, and λ is a local prime element lying above 2. The λ-adic expansion of an element of o is then of type

$$a + b\lambda + c\lambda^2 + d\lambda^3 \pmod{4o},$$

where a, b, c, d = 0, 1. On the other hand, we say that the ramification is of second kind if $\text{Tr}(\lambda) \equiv 0 \pmod{4}$ for all local prime elements λ lying above 2. The table of values for F_4 is then as follows.

	$F_4(0)$	$F_4(1)$	$F_4(2)$	$F_4(3)$
$\left(\frac{k}{2}\right) = 1$	2	0	2	0
$\left(\frac{k}{2}\right) = -1$	2/3	4/3	2/3	4/3
$\left(\frac{k}{2}\right) = 0$ First kind	2	0	2	0
$\left(\frac{k}{2}\right) = 0$ Second kind	0	0	4	0

We leave the computations as an exercise.

THE MODEL FOR THE MIXED CASE

§3. The mixed Galois representation

Let
$$W = \prod_\ell GL_2(\mathbb{Z}_\ell).$$

We let K be an infinite Galois extension of $\mathbb{Q}$, let $G = \text{Gal}(K/k)$, and let
$$\rho' : G \longrightarrow \prod GL_2(\mathbb{Z}_\ell) = W$$

be a GL_2-representation (open embedding) as in Part I. We write ρ' instead of ρ because we shall use ρ'' for a representation related to k, and then we put the two together. As in Part I, we assume that there is a positive integer Δ such that the ℓ-adic representation ρ'_ℓ is unramified at p if $p \nmid \ell\Delta$. The Frobenius element σ'_p then has a characteristic polynomial

$$X^2 - t_p X + p,$$

assumed independent of ℓ, and with integer trace t_p. If ρ' is the representation associated with an elliptic curve, then Δ is the discriminant of the curve. Finally, since we do not explicitly assume that ρ' comes from an elliptic curve, we do assume the Riemann hypothesis that the roots of the characteristic polynomial have absolute value $\sqrt{p}$.

If L is a set of primes, we denote by G_L the projection of G under the representation
$$\rho'_L : G \longrightarrow \prod_{\ell \in L} GL_2(\mathbb{Z}_\ell).$$

We may write
$$\rho'_L = \prod_{\ell \in L} \rho'_\ell.$$

We let K_L be the fixed field of the kernel of ρ'_L, so that
$$G_L = \text{Gal}(K_L/\mathbb{Q}).$$

If M is a positive integer, and L is the set of primes dividing M, then we also write

$$G_M = G_L \quad \text{and} \quad K_M = K_L .$$

On the other hand, if L is set of primes complementary to the primes dividing M, then we use the notation

$$G_L = G_{[M]} .$$

In the preceding section, we had described some facts concerning the class field theory above k. We now mix k_{ab} and the GL_2-extension K. The relevant lattice of fields is illustrated.

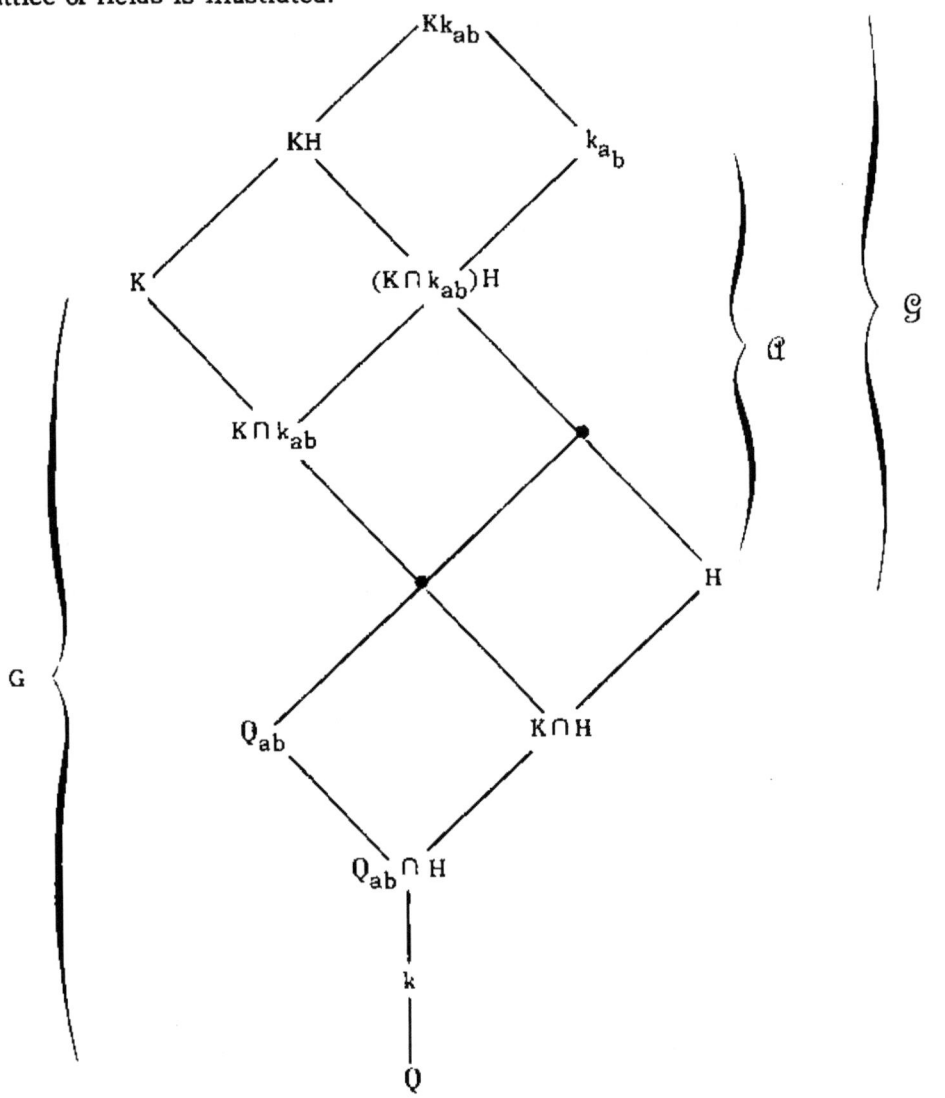

We shall consider the compositum $K = Kk_{ab}$, and let

$$\mathcal{G} = \text{Gal}(Kk_{ab}/H).$$

Then we have an embedding $\rho = (\rho', \rho'')$,

$$\rho : \mathcal{G} \longrightarrow G \times \mathfrak{A} \longrightarrow \prod GL_2(Z_\ell) \times \left(\prod \mathfrak{o}_\ell^*\right)/\mathfrak{o}^* = W \times \bar{U}.$$

The image of ρ is contained in the subgroup of the product consisting of those elements (σ', σ'') which have the same effect on the cyclotomic field Q_{ab}, which is contained in both K and k_{ab}. If $\sigma \in \mathcal{G}$, then σ' is its restriction to K and σ'' is its restriction to k_{ab}. We denote by a subscript N under the cross sign the fibering with respect to the effect on the roots of unity. We may therefore say that $\mathcal{G}$ is contained in

$$G \times_N \mathfrak{A}.$$

In terms of the matrix representations ρ' and ρ'', this means that the image of ρ is contained in the group of elements (a, u) such that

$$\det a = Nu,$$

where

$$\det : \prod GL_2(Z_\ell) \longrightarrow \prod Z_\ell^*$$

$$N : \prod \mathfrak{o}_\ell^* \longrightarrow \prod Z_\ell^*$$

are the determinant and norm maps respectively. The group of such matrices is denoted by the fibering notation

$$W \times_N \bar{U} = \prod GL_2(Z_\ell) \times_N \left(\prod \mathfrak{o}_\ell^*\right)/\mathfrak{o}^*.$$

We let $S\mathfrak{o}_\ell^*$ be the subgroup of $\mathfrak{o}_\ell^*$ consisting of those elements with norm 1 (the special linear group). Similarly, SG is the subgroup of G consisting of elements with determinant 1, and $S\mathcal{G}$ is the subgroup of $\mathcal{G}$ which fixes Q_{ab}. We have a pair of exact sequences:

$$\begin{array}{ccccccccc}
1 & \longrightarrow & SW \times S\bar{U} & \longrightarrow & W \times_N \bar{U} & \longrightarrow & \prod Z_\ell^* & \longrightarrow & 1 \\
& & \uparrow & & \uparrow & & \uparrow & & \\
1 & \longrightarrow & S\mathcal{G} & \longrightarrow & \mathcal{G} & \longrightarrow & \det \mathcal{G} & \longrightarrow & 1
\end{array}$$

and the bottom sequence maps injectively into the top one. The image of $S\mathcal{G}$ in

$$SW \times S\bar{U} = \prod SL_2(Z_\ell) \times \left(\prod S\mathfrak{o}_\ell^*\right)/\mathfrak{o}^*$$

is of finite index, and similarly the image of $\mathcal{G}/S\mathcal{G}$ in $\prod Z_\ell^*$ is of finite index. Hence the image of $\mathcal{G}$ in the fiber product is of finite index, it is closed and hence open.

When dealing with these fiber products, it is convenient to get rid of the roots of unity, and to lift the Galois group as follows. We have a w to 1 covering

$$\prod GL_2(Z_\ell) \times_N \prod \mathfrak{o}_\ell^* \longrightarrow \prod GL_2(Z_\ell) \times_N \bar{U}.$$

We let $\tilde{\mathcal{G}}$ be the inverse image of $\mathcal{G}$ in this covering, so we have a commutative diagram:

$$\begin{array}{ccc} \tilde{\mathcal{G}} & \longrightarrow & \prod GL_2(Z_\ell) \times_N \prod \mathfrak{o}_\ell^* \\ \downarrow & & \downarrow \\ \mathcal{G} & \longrightarrow & \prod GL_2(Z_\ell) \times_N \left(\prod \mathfrak{o}_\ell^*\right)/\mathfrak{o}^* \end{array}$$

Similarly, we can define $\tilde{\mathcal{A}}$ and we can identify

$$\tilde{\mathcal{A}} = U.$$

The rest of this section is devoted to two separate topics. First an analysis of the intersection $K \cap k_{ab}$. The reader may skip this, since it plays no role in the general determination of the desired constant. It is of course important to have for the determination in special cases.

Second, we discuss the reduction of the Galois group to finite levels. The reader should glance briefly at the definitions and then read into §4 immediately, referring to the formal development only if he needs it especially.

The intersection $K \cap k_{ab}$

This intersection will play a crucial role in the determination of the coincidence relations between Frobenius elements in G and Frobenius elements in k_{ab}.

We let $G_k = \text{Gal}(K/k)$. Then

$$[K \cap k_{ab} : Q_{ab}] = (G' : G'_k),$$

where the prime superscript indicates the closure of the commutator subgroup, cf. Part I. It is easy to see that the index on the right is finite, and it will follow from stronger lemmas to be proved below, concerning commutator subgroups of open subgroups of $GL_2(Z_\ell)$. By convention, when we speak of a commutator subgroup, we shall always mean the closure of the subgroup generated by commutators. We let $M_\ell = \text{Mat}_2(Z_\ell)$.

We begin with a lemma concerning the "generic" case, just to see what happens almost always.

Lemma 1. *Let q be a prime ≥ 5. Then*

$$GL_2(Z_q)' = SL_2(Z_q) = SL_2(Z_q)'.$$

Proof. Clearly, $GL_2(Z_q)'$ is a subgroup of $SL_2(Z_q)$, and is closed. For $q \geq 5$ it is standard finite group theory that

$$SL_2(F_q)' = SL_2(F_q).$$

Hence $SL_2(Z_q)'$ is a closed subgroup which reduces modulo q to $SL_2(F_q)$. The refinement lemma of Serre shows that $SL_2(Z_q)'$ must be equal to $SL_2(Z_q)$ and proves what we want.

If V is any open subgroup of $SL_2(Z_q)$ for any prime q, then V' is obviously open. This already makes it obvious that

$$(G' : G'_k)$$

is finite.

Theorem 3.1. *Let q be a prime ≥ 5 which divides D. Assume that*

$$G = GL_2(Z_q) \times G_L,$$

where L is the complement of $\{q\}$. Then

$$K \cap k_{ab} = Q_{ab}.$$

Proof. We have $G_q = GL_2(Z_q)$, and G_k is of index 2 in G. The group G_q has a unique subgroup of index 2, the subgroup E_q consisting of all elements σ such that

$$\det \sigma \in Z_q^{*2}.$$

Hence $E_q \times \{e_L\}$ is contained in G_k. Note that E_q contains $SL_2(Z_q)$. But

$$SL_2(Z_q)' = SL_2(Z_q)$$

by Lemma 1. Consequently $E_q' \supset SL_2(Z_q)$, and we find that

$$G_k' \supset SL_2(Z_q) \times \{e_L\}.$$

On the other hand, k cannot be contained in K_L because K_L is disjoint from K_q over the rationals, so that if $k \subset K_L$ then k would be contained in a cyclotomic field disjoint from the q^n-th roots of unity, whence would be unramified at q, which contradicts our hypothesis. If pr_L denotes the projection on the L-th factor, we obtain

$$pr_L G_k = G_L,$$

and therefore

$$pr_L G_k' = G_L',$$

Since $G_k' \supset SL_2(Z_q) \times \{e_L\}$, we conclude that

$$G_k' \supset SL_2(Z_q) \times G_L' = G',$$

thereby proving that $G_k' = G'$, and also proving the theorem.

We recall that

$$\mathcal{G} = Gal(Kk_{ab}/H) \quad \text{and} \quad \mathcal{A} = Gal(k_{ab}/H).$$

We let

$$G \times_N \mathcal{A}$$

be the set of pairs $(\sigma', \sigma'') \in G \times \mathcal{A}$ such that σ' and σ'' have the same effect on Q_{ab}.

Theorem 3.2. *Assume that* $K \cap k_{ab} = Q_{ab}$. *Then*

$$\mathcal{G} = G \times_N \mathcal{A}.$$

Proof. The situation is illustrated by the following diagram,

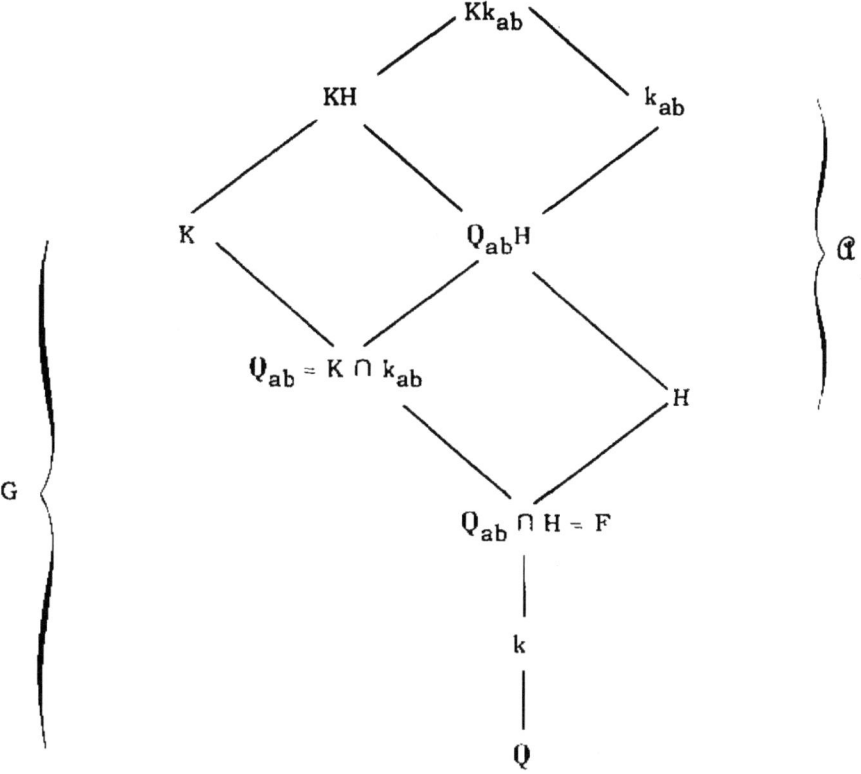

We let $F = Q_{ab} \cap H$. Given $\sigma' \in G$ and $\sigma'' \in \mathcal{A}$ having the same effect on Q_{ab}, we have to show there exists $\sigma \in \mathcal{G}$ which restricts to σ' on K and σ'' on k_{ab}. Observe first that σ' is trivial on F. From the diagram, we can lift σ' to an element of Gal(KH/H), since H is disjoint from K over F. The element σ'' then coincides with σ' on $Q_{ab}H$. The two extensions KH and k_{ab} of

$Q_{ab}H$ are disjoint over $Q_{ab}H$. We can therefore find an automorphism σ of Kk_{ab} which gives σ'' on KH and σ' on k_{ab}, as desired.

In some applications, it is convenient to have a refinement of the above statement, taking into account only part of the Galois group. Let L be a set of primes We may put L as an index in the preceding theorem in the following manner. We have the group G_L, which is the projection of G in the partial product

$$\prod_{\ell \in L} GL_2(Z_\ell),$$

and G_L is the Galois group of the field denoted by K_L. Similarly, we let

$$\mathcal{Q}_L = \left(\prod_{\ell \in L} \mathfrak{o}_\ell^* \right) / \mathfrak{o}^*,$$

so that $\mathcal{Q}_L$ is the Galois group of an extension of H which we denote by k_L^{ab}. We let

$$\mathcal{G}_L = \text{Gal}(K_L k_L^{ab}/H)$$

be the group of the mixed extension. We have the cyclotomic field Q_L^{ab} whose Galois group is

$$\prod_{\ell \in L} Z_\ell^*.$$

Theorem 3.2L. *Assume that* $K_L \cap k_L^{ab} = Q_L^{ab}$. *Then*

$$\mathcal{G}_L = G_L \times_N \mathcal{Q}_L.$$

Proof. The same as before, mutatis mutandis.

Under the representations ρ' and ρ'', the preceding theorem shows that

$$\tilde{\mathcal{G}} = \rho'(G) \times_N \prod_\ell \mathfrak{o}_\ell^*,$$

or omitting the ρ' for simplicity,

$$\tilde{\mathcal{G}} = G \times_N \prod_\ell \mathfrak{o}_\ell^*.$$

Example. Suppose that G is Serre's subgroup, cf. §10, below, or Part I,

$$G = S_{2q} \times \prod_{\ell \neq 2, q} GL_2(Z_\ell).$$

Let us define

$$\tilde{S}_{2q} = S_{2q} \times_N o^*_{2q}.$$

Theorem 3.3. *Assume that G is Serre's subgroup as above. Assume also that* $K \cap k_{ab} = Q_{ab}$. *Then*

$$\tilde{\mathcal{G}} = \tilde{S}_{2q} \times \prod_{\ell \neq 2, q} [GL_2(Z_\ell) \times_N o^*_\ell].$$

Proof. Special case of the previous theorem.

We shall now state the results to be proved in Part III concerning the intersection $K \cap k_{ab}$.

Theorem 3.4. *If* $D = -8, -24, -15, -20, -40, -55, -88$ *then* $K \cap k_{ab} = Q_{ab}$ *for all of our five curves*

A, B, C, D, $X_0(11)$.

Proof. The case -8 is proved in Part III, Theorem 3.2, the case -24 in Theorem 3.3. The other cases are proved in Theorems 3.4, 3.5, 3.6.

Next we give a table of intersections, also giving the reference to the theorem in Part III which proves the assertion.

	$K \cap k_{ab}$
-3	$Q_{ab}(\Delta^{\frac{1}{3}})$, all curves except D, by 4.2
-4	$Q_{ab}(\Delta^{\frac{1}{4}})$, all curves, by 5.2
-43	$Q_{ab}(B_2, \Delta^{\frac{1}{4}})$ for curve B, by 5.1
-11	$Q_{ab}(X_0(11)_2, \Delta^{\frac{1}{4}})$ for $X_0(11)$, by 5.1
-3	$Q_{ab}(D_2, \Delta^{\frac{1}{3}}, \Delta^{\frac{1}{4}})$ for curve D, by 5.1

Reduction mod M

Let M be a positive integer. For each ℓ we let $W_{\ell,M}$ consist of those matrices $a \in GL_2(\mathbb{Z}_\ell)$ such that
$$a \equiv I \pmod{M}.$$
If $\ell \nmid M$ then this condition is empty. If $\ell \mid M$, then it means the usual congruence mod M $Mat_2(\mathbb{Z}_\ell)$.

Similarly, we let $U_{\ell,M}$ be the subgroup of elements u of $\mathfrak{o}_\ell^*$ such that $u \equiv 1 \pmod{M\mathfrak{o}_\ell}$.

We let
$$W_M = \prod_\ell W_{\ell,M} \quad \text{and} \quad U_M = \prod_\ell U_{\ell,M}.$$

If $M \geq 3$ then $U_M \cap \mathfrak{o}^* = \{1\}$. In this case we may view U_M as a subgroup of
$$\left(\prod \mathfrak{o}_\ell^*\right)/\mathfrak{o}^*.$$

Since $\rho(\mathcal{G})$ is open in the fiber product
$$\prod GL_2(\mathbb{Z}_\ell) \times_N \left(\prod \mathfrak{o}_\ell^*\right)/\mathfrak{o}^*,$$
there exists an integer M_0 having the following property.

If $M_0 | M$, then $\rho(\mathcal{G})$ contains the fiber product
$$W_M \times_N \bar{U}_M.$$

Such an integer M_0 will be said to **stabilize** ρ, and we let
$$\mathcal{G}_M = \rho^{-1}(W_M \times_N \bar{U}_M).$$

We often identify $\mathcal{G}_M$ with its matrix representation
$$\rho(\mathcal{G}_M) = W_M \times_N \bar{U}_M.$$

If we don't, then we write $\sigma = \sigma(a, u)$ to mean $(a, u) = \rho(\sigma)$. We let $\mathcal{G}(M) = \mathcal{G}/\mathcal{G}_M$.

Similarly, we let G_M be the subgroup of G corresponding to W_M, i.e. such that $\rho'(G_M) = W_M$, and $\mathcal{A}_M$ be the subgroup of $\mathcal{A}$ such that $\rho''(\mathcal{A}_M) = U_M$. We let

$$G(M) = G/G_M \quad \text{and} \quad \mathcal{A}(M) = \mathcal{A}/\mathcal{A}_M .$$

We let $K(M)$ and $k_{ab}(M)$ be the fixed fields of G_M and $\mathcal{A}_M$ respectively. We then get:

The fixed field of G_M is $K(M) k_{ab}(M)$.

We already had

$$\rho'_{(M)} : \mathrm{Gal}(K/\mathbb{Q}) = G \longrightarrow GL_2(M) = W(M) .$$

We therefore obtain

$$\rho_{(M)} : \mathcal{G} \longrightarrow W(M) \times_N \bar{U}(M) .$$

We let

$$\mathcal{K} = K k_{ab} \quad \text{and} \quad \mathcal{K}(M) = K(M) k_{ab}(M) ,$$

so that

$$\mathcal{G}(M) = \mathrm{Gal}(\mathcal{K}(M)/H) .$$

The lattice of fields and groups may be drawn as follows.

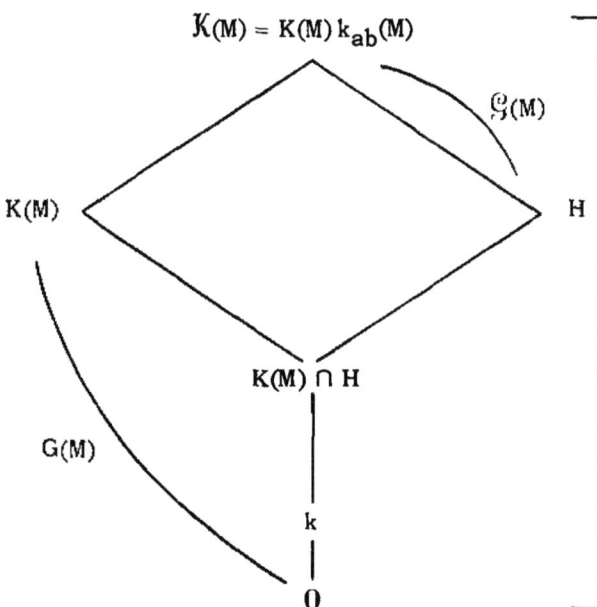

We let r_M denote reduction mod M. If M_0 stabilizes ρ, and $M_0|M$, then we have a commutative diagram:

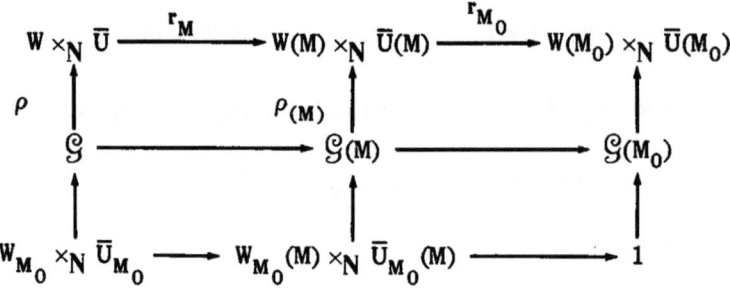

When we write the N as an index to the cross sign in a fiber product mod M, say in
$$W_{M_0}(M) \times_N \bar{U}_{M_0}(M) ,$$
the fibering is of course to be interpreted mod M. Elements of this fibered product are pairs $(\bar{a}, \bar{u})$ such that
$$\det \bar{a} \equiv N\bar{u} \pmod{M} .$$

Remark. *Given such a pair $(\bar{a}, \bar{u})$ with $\det \bar{a} = N\bar{u}$ mod M, there exists* $(a, u) \in W_{M_0} \times_N \bar{U}_{M_0}$ *such that*
$$r_M(a, u) = (\bar{a}, \bar{u}) .$$

This is obvious by solving a linear equation whose leading coefficient is a unit. In particular, the bottom row in our diagram is exact.

We can also express the stabilizing condition in a form analogous to that which we used previously for the simpler supersingular case.

If M_0 stabilizes ρ, and $M_0|M$, then $\mathcal{G}(M)$ is the inverse image of $\mathcal{G}(M_0)$ in
$$W(M) \times_N \bar{U}(M) ,$$
under the reduction map r_{M_0}. Similarly, $\mathcal{G}$ is the inverse image of $\mathcal{G}(M_0)$ in
$$W \times_N \bar{U} .$$

In the above statement, we identified $\mathcal{G}$ with its image in the corresponding matrix groups. We let $\tilde{\mathcal{G}}_M$ be the projection on the factors with $\ell|M$ as usual.

We say that M splits $\rho = (\rho', \rho'')$ if $\tilde{\mathcal{G}}$ has an expression

$$\tilde{\mathcal{G}} = \tilde{\mathcal{G}}_M \times \prod_{\ell \nmid M} [GL_2(\mathbb{Z}_\ell) \times_N \mathfrak{o}_\ell^*].$$

The above discussion shows that there always exists an integer M which splits and stabilizes ρ.

§4. The probabilistic model

We fix a positive integer Δ such that, for all M, if $\mathfrak{p} \nmid M\Delta$, then p is unramified in K(M) (that is, $\rho_{(M)}$ is unramified at p). If $\mathfrak{p} \nmid M\Delta D$, then p is unramified in $\tilde{K}(M) = K(M) k_{ab}(M)$. When ρ is the representation associated with an elliptic curve, then Δ is the discriminant.

Let M be a positive integer ≥ 3 and let t, s be integers. Let $\mathfrak{p}$ be a prime ideal above p in k. We define:

$$r_M(\mathfrak{p}, t, s) = \begin{cases} 0 & \text{unless } \mathfrak{p} \nmid M\Delta D, \; \text{tr}\, \rho'_{(M)}(\sigma'_\mathfrak{p}) \equiv t \bmod M, \\ & \mathfrak{p} \text{ splits completely and its factors are principal in } k. \\ \\ & \text{otherwise, number of generators } \pi \text{ of } \mathfrak{p} \\ & \text{such that } \text{Tr}\, \pi \equiv s \bmod M. \end{cases}$$

We note that $r_M(\mathfrak{p}, t, s) = 0$ unless $\text{tr}\, \rho'_{(M)}(\sigma'_\mathfrak{p}) \equiv t \pmod{M}$, and also there is an element in $\rho''_{(M)}(\sigma''_\mathfrak{p})$ whose trace is $\equiv s \pmod{M}$. The function r_M counts the multiplicity of such elements in that situation. We define $F_M(t, s)$ by the condition:

$$\frac{w}{M^2} F_M(t, s) = \lim_{x \to \infty} \frac{1}{|P_k(x)|} \sum_{\mathfrak{p} \in P_k(x)} r_M(\mathfrak{p}, t, s).$$

Thus, roughly speaking, the limit on the right is the density relative to P_k of those primes $\mathfrak{p} \in P_k$ such that $t \equiv \text{tr}\, \rho'_{(M)}(\sigma'_\mathfrak{p})$, and such that $s \equiv \text{Tr}\, \rho''_{(M)}(\sigma''_\mathfrak{p})$, except that this last congruence has to be explained accurately by means of the function $r_M(\mathfrak{p}, t, s)$. Having defined F_M in this manner, we then have

(1) $$\sum_{(t,s) \bmod M} \frac{1}{M^2} F_M(t,s) = 1.$$

We remind the reader that $\mathfrak{G}(M) = \text{Gal}(K(M) k_{ab}(M)/\mathbb{Q})$ but

$$\mathcal{G}(M) = \text{Gal}(K(M) k_{ab}(M)/H).$$

Thus elements of $\mathcal{G}(M)$ leave H fixed, and if $\sigma \in \mathcal{G}(M)$ then $\sigma'' \in \mathfrak{A}(M)$. Cf. the diagram of §3.

We define a function $\lambda_M(\sigma, t, s)$ for any pair of integers t, s and $\sigma \in \text{Gal}(\overline{K}(M)/\mathbb{Q}) = \mathfrak{G}(M)$ by:

$$\lambda_M(\sigma, t, s) = \begin{cases} 0 & \text{unless } \sigma \in \mathcal{G}(M) \text{ and } \text{tr}\, \rho'_{(M)}(\sigma') \equiv t \bmod M; \\ & \text{otherwise, number of elements in } \rho''_{(M)}(\sigma'') \\ & \text{whose trace is } \equiv s \bmod M. \end{cases}$$

We view $\rho''_{(M)}(\sigma'')$ as a coset πo^* in $o(M)^*/o^*$. The definition is designed to transfer our counting problem to the Galois group, because if $p \nmid M\Delta D$, then

$$\lambda_M(\sigma_p, t, s) = r_M(p, t, s).$$

It is also useful to deal with a lifting of the Galois group which gets rid of o^*. We have a w to 1 covering

$$G(M) \times o(M)^* \longrightarrow G(M) \times o(M)^*/o^*$$

and we let $\tilde{\mathcal{G}}(M)$ be the inverse image of $\mathcal{G}(M)$ in this covering, so we have a commutative diagram:

$$\begin{array}{ccc} \tilde{\mathcal{G}}(M) & \longrightarrow & G(M) \times o(M)^* \\ \downarrow & & \downarrow \\ \mathcal{G}(M) & \longrightarrow & G(M) \times o(M)^*/o^*. \end{array}$$

We let $\tilde{\mathcal{G}}(M)_{s,t}$ = subset of elements σ in $\mathcal{G}(M)$ such that

$$\text{tr}\, \sigma' \equiv t \quad \text{and} \quad \text{Tr}\, \sigma'' \equiv s \bmod M.$$

Theorem 4.1. *If* $M \geq 3$ *then*

$$\frac{1}{M^2} F_M(t,s) = \frac{|\tilde{\mathcal{G}}(M)_{t,s}|}{|\tilde{\mathcal{G}}(M)|}.$$

Proof. The index of $\mathcal{G}(M)$ in $\mathfrak{S}(M)$ is precisely $2h$, where h is the class number ($= (H:k)$). Thus from the definition, we obtain

$$\frac{w}{M^2} F_M(t,s) = \lim_{x \to \infty} \frac{2h}{|P(x)|} \sum_{p \in P(x)} \lambda_M(\sigma_p, t, s)$$

$$= \frac{2h}{|\mathfrak{S}(M)|} \sum_{\sigma \in \mathfrak{S}(M)} \lambda_M(\sigma, t, s) \quad \text{(by Tchebotarev)}$$

$$= \frac{1}{|\mathcal{G}(M)|} \sum_{\sigma \in \mathcal{G}(M)} \lambda_M(\sigma, t, s)$$

$$= w \frac{|\mathcal{G}(M)_{t,s}|}{|\mathcal{G}(M)|} ,$$

thereby proving the theorem.

We have fixed the interval J (corresponding to a sector of width π/w) and a positive integer M which splits ρ''. For each prime p, the fiber of the probabilistic model at p is now $Z \times Z$. The measure $\mu_{p,J}$ is represented with respect to counting measure by a function

$$f_{M,J}(t, s, p) \geq 0 ,$$

which we assume of the form

PR 3. $\qquad f_{M,J}(t, s, p) = c_p\, g(\xi(t, p))\, g''_J(\xi(s, p))\, F_M(t, s) ,$

where c_p is a constant such that

(2) $$\sum_{(t,s)} f_{M,J}(t, s, p) = 1 .$$

We can now prove successively that $c_p \sim 1/4p$, and that the above assumption implies the Tchebotarev, Sato-Tate, and Hecke distribution properties. Thus again, we have picked a simple probabilistic model compatible with these properties. Note that **PR 3** amounts to a condition of independence for the behaviors at infinity (in the GL_2 extension and the imaginary quadratic field) from each other, and from the behavior under congruence conditions.

We give the proof that $c_p \sim 1/4p$ in full detail to show the reader how the somewhat more complicated system under which we now work, nevertheless parallels the previous one quite closely.

We start with the integral

$$1 = \mu_{g''}(J) = \int_{-1}^{1} \int_{-1}^{1} g(\xi) g''_J(\xi'') \, d\xi \, d\xi''.$$

We have a product decomposition of the double sum,

$$\frac{1}{4p} \sum_{(t,s)} g(\xi(t,p)) g''_J(\xi(s,p)) = \left(\frac{1}{2\sqrt{p}} \sum_t g(\xi(t,p)) \right) \left(\frac{1}{2\sqrt{p}} \sum_s g''_J(\xi(s,p)) \right).$$

The first factor is an approximating Riemann sum for the constant 1. The second factor is a lower sum, which approaches 1 as $p \to \infty$. Therefore the double sum on the left approaches 1 also. The double sum, summed only for pairs $(t, s) \equiv (t_0, s_0) \mod M$ approaches

$$\frac{1}{M^2}.$$

Multiplying with $F_M(t_0, s_0)$ we obtain the limit

(3) $\displaystyle \lim_p \frac{1}{4p} \sum_{(t,s) \equiv (t_0, s_0) \bmod M} g(\xi(t,p)) g''_J(\xi(s,p)) F_M(t_0, s_0) = \frac{1}{M^2} F_M(t_0, s_0).$

Summing over all congruence classes $(t_0, s_0) \mod M$ yields the value 1. Comparing with the definition of $f_{M,J}(t, s, p)$ and the normalization of c_p concludes the proof that

$$c_p \sim 1/4p.$$

We have reached an analogous point to that reached in the supersingular case, where we can have lemmas corresponding to Lemma 1 and Lemma 2, to show compatibility of our probability model with Tchebotarev, Sato-Tate and Hecke for random sequences, using the law of large numbers. The routine is the same, and will be omitted, since it is not used in the sequel.

§5. The asymptotic behavior

We finally come to counting what we want, which amounts to computing the measure of the diagonal D in each fiber over each prime. We have

$$\mu_{p,J}(D) = \sum_t f_{M,J}(t,t,p)$$

$$= \sum_t c_p\, g(\xi(t,p))\, g''_J(\xi(t,p))\, F_M(t,t)$$

$$= \sum_{t_0 \bmod M} F_M(t_0, t_0) \sum_{t \equiv t_0} c_p\, g(\xi(t,p))\, g''_J(\xi(t,p))$$

The sum over $t \equiv t_0$ on the right is a Riemann sum, which is asymptotic to

$$\frac{1}{2\sqrt{p}} \frac{1}{M} \int_{-1}^{1} g(\xi)\, g''_J(\xi)\, d\xi \, .$$

We abbreviate

$$\boxed{\int_{-1}^{1} g(\xi)\, g''_J(\xi)\, d\xi = C_J^{\infty}(k,\rho) \, .}$$

It is the part at infinity for our constant, relative to the interval J. We then find

$$\mu_{p,J}(D) \sim C_J^{\infty}(k,\rho) \sum_{t_0 \bmod M} \frac{1}{M} F_M(t_0, t_0) \cdot \frac{1}{2\sqrt{p}} \, .$$

It follows from §7, Lemma 1 and Theorem 9.1 that the limit over suitable M exists:

$$\boxed{\lim_M \sum_{t_0 \bmod M} \frac{1}{M} F_M(t_0, t_0) = C^{\text{fin}}(k,\rho) \, ,}$$

defining the finite part of the desired constant. We then define

$$C_J(k,\rho) = \frac{1}{2h} C_J^\infty(k,\rho) C^{fin}(k,\rho).$$

We have reintroduced a factor $1/2h$ because originally, we determined F_M by the density of primes relative to P_k, but we want at the very end to have the density relative to the whole set of primes. We then define the total constant, summing over w disjoint intervals J covering $[-1, 1]$:

$$C(k,\rho) = \sum_J C_J(k,\rho) = \frac{1}{2h} C^\infty(k,\rho) C^{fin}(k,\rho),$$

where

$$C^\infty(k,\rho) = \int_{-1}^{1} \frac{wg(\xi)}{\pi\sqrt{1-\xi^2}} d\xi.$$

The conjecture is that

$$N_{k,\rho}(x) \sim C(k,\rho)\pi_{\frac{1}{2}}(x).$$

Remark. When the Sato-Tate function is given by $\phi(\theta) = \frac{2}{\pi}\sin^2\theta$, or in other words

$$g(\xi) = \frac{2}{\pi}\sqrt{1-\xi^2},$$

then we get a rather convenient unexpected cancellation, and the infinity part of the constant is given by

$$C_J^\infty(k,\rho) = \frac{2w}{\pi^2} \text{ (length of J)}.$$

Thus

$$C^\infty(k,\rho) = 4w/\pi^2$$

and we find

$$C(k,\rho) = \frac{1}{2h} \frac{4w}{\pi^2} C^{fin}(k,\rho).$$

It will be proved in Theorem 6.3 that the finite part of the constant has a product decomposition

$$C^{fin}(k,\rho) = C_M \prod_{\ell \nmid M} C_\ell$$

for a suitable integer M. Furthermore, the non-special factors C_ℓ will be computed and tabulated in §9, showing that if

$$L(1,\chi)_\ell = \left(1 - \left(\frac{k}{\ell}\right)\frac{1}{\ell}\right)^{-1}$$

is the ℓ-component of the product expression for the non-trivial series of k at 1, then C_ℓ differs from this ℓ-component by a factor $O(1+1/\ell^2)$. Consequently it is useful to make a further transformation of the expression for the constant, and to give a name to these factors, which give rise to an absolutely convergent product.

For $\ell \nmid M$ we let

$$C'_\ell = \begin{cases} 1 & \text{if } \left(\frac{k}{\ell}\right) = 1 \\ 1 + \dfrac{2r^2}{(1-r^2)(1+r)} & \text{if } \left(\frac{k}{\ell}\right) = -1 \\ \dfrac{1}{1-r^2} & \text{if } \left(\frac{k}{\ell}\right) = 0 \end{cases}$$

Recall that $r = 1/\ell$. Then Theorem 9.1 will show that

$$C_\ell = L(1,\chi)_\ell C'_\ell .$$

We define C'_M by the relation

$$C_M = C'_M \prod_{\ell \mid M} L(1,\chi)_\ell .$$

Then

$$C^{\text{fin}}(k,\rho) = L(1,\chi) C'_M \prod_{\ell \nmid M} C'_\ell .$$

For the quadratic field k we have the formula

$$L(1,\chi) = \frac{2\pi h}{w\sqrt{|D|}} .$$

Putting all this together yields:

Theorem 5.1. *We have the product decomposition*

$$C(k,\rho) = \frac{4}{\pi\sqrt{|D|}} C'_M \prod_{\ell \nmid M} C'_\ell .$$

As mentioned previously, the infinite product in this expression is absolutely convergent. The value $1/\sqrt{|D|}$ is of course known. Hence the computation of the constant is easily reduced to the exceptional factor C'_M, which depends on the part of the Galois group which cannot be easily split or decomposed, and has to be studied in special cases separately, depending on their idiosyncrasies. This factor C'_M is a quotient of C_M by a finite product of local L-series terms $L(1,\chi)_\ell$ for $\ell | M$. We shall see that C_M is determined by a singular measure on the M-component of a Galois group. Cf. Theorem 6.3.

§6. The finite part of the constant as a quotient of integrals

The finite part of the constant is expressed as a limit over M. We describe how it can be interpreted as an integral, first at finite level.

Let M denote Mat_2, so that $M(M) = \text{Mat}_2(M) = \text{Mat}_2(Z/MZ)$. We let

$$T' = (\text{tr}, \det)$$

be the map sending a matrix to its trace and determinant. At finite level, we should write

$$T'_{(M)} : M(M) \longrightarrow Z(M)^2 \;,$$

but we omit the subscript (M) and write simply T when M is fixed. Similarly, we have the trace-norm map

$$T'' = (\text{Tr}, N)$$

on elements of $\mathfrak{o}$, whence

$$T''_{(M)} : \mathfrak{o}(M) \longrightarrow Z(M)^2 \;.$$

Haar measure on any compact group, and especially on a finite group, will be assumed to be normalized to give the group measure 1, unless otherwise specified. Here we have in mind M(M), Z(M), and $\mathfrak{o}$(M).

Similarly, we have the M-adic corresponding notions, namely:

$$M_\ell = \text{Mat}_2(Z_\ell), \qquad M_M = \text{Mat}_2(Z_M) = \prod_{\ell \mid M} M_\ell$$

$$T' = T'_M : M_M \longrightarrow Z_M^2$$

on the GL_2 side, and

$$T'' = T''_M : \mathfrak{o}_M \longrightarrow Z_M^2$$

on the quadratic field side, where

$$\mathfrak{o}_M = \prod_{\ell \mid M} \mathfrak{o}_\ell \;.$$

Let S be an open subset of M_M. Then $S(M) \subset M(M)$ is the reduction of S mod M.

We let μ denote Haar measure, with a subscript to indicate the corresponding group. If S is an open subset of the group, we let μ_S denote the restriction of Haar measure to S, and 0 on the complement of S. It will be proved in the next sections that the map T' has a continuous density function with respect to Haar measures. It is essentially clear from the computations of §2 that the similar map T'' on the quadratic field side also has a continuous density function. These two functions are denoted by h' and h'' respectively, although we sometimes write h instead of h'. By definition, we have the expression for the direct image of Haar measure M_M and o_M respectively; with $X \subset M_M$ and $Y \subset o_M$:

$$T'_* d\mu_X = h'_X d\mu_{Z_M^2} \quad \text{and} \quad T''_* d\mu_Y = h''_Y d\mu_{Z_M^2}.$$

We shall deal with subsets S of the fibered product

$$S \subset M_M \times_N o_M.$$

Such a subset is said to decompose at level M if it is a finite disjoint union of fibered products

$$S = \bigcup S_i = \bigcup X_i \times_N Y_i, \qquad S_i = X_i \times_N Y_i,$$

where X_i is open in M_M, Y_i is open in o_M, and X_i, Y_i are stable at level M (in other words, are the inverse images of their reductions mod M).

For such a subset S, we define

$S(M)_{t,t}$ = set of elements $(g,a) \in S(M)$ such that $\operatorname{tr} g = \operatorname{Tr} a \equiv t$.

The equality on the right is to be viewed in $Z(M)$, i.e. as a congruence mod M. We define

$$C_{(M)}(S) = M \sum_{t \bmod M} \frac{|S(M)_{t,t}|}{|S(M)|}.$$

It is convenient to normalize the numerator and denominator of this expression by multiplying with M^5. Thus we let

$$M^5 \operatorname{Num}_{(M)}(S) = M \cdot \#\{(g,a) \in S(M), \operatorname{tr} g = \operatorname{Tr} a\}.$$

The denominator is just
$$M^5 \operatorname{Den}_{(M)}(S) = |S(M)|.$$

We shall write the numerator and denominator as integrals, using the density functions $h'_{X(M)}$ and $h''_{Y(M)}$, thereby giving an expression for $C_{(M)}(S)$ as a quotient of integrals,
$$C_{(M)}(S) = \operatorname{Num}_{(M)}(S)/\operatorname{Den}_{(M)}(S).$$

Theorem 6.1. *We have:*
$$\operatorname{Num}_{(M)}(S) = \sum_i \iint_{Z(M)^2} h'_{X_i(M)}(t,u) h''_{Y_i(M)}(t,u)\,du\,dt$$

$$\operatorname{Den}_{(M)}(S) = \sum_i \iiint_{Z(M)^3} h'_{X_i(M)}(t',u) h''_{Y_i(M)}(t'',u)\,du\,dt'\,dt''$$

Proof. Without loss of generality, we may assume that
$$S = X \times_N Y.$$

For simplicity of notation, we write S, X, Y instead of $S(M)$, $X(M)$ and $Y(M)$ respectively, so we work entirely under reduction mod M. Then the numerator is given by

$$M^5 \operatorname{Num}_{(M)}(S) = M \#\{(g,a) \in X \times Y,\ \operatorname{tr} g = \operatorname{tr} a,\ \det g = Na\}$$

$$= M \sum_{(t,u)} \#\{(g,a) \in X \times Y,\ \operatorname{tr} g = \operatorname{Tr} a = t,\ \det g = Na = u\}$$

$$= M \sum_{(t,u)} \#\{g \in X,\ \operatorname{tr} g = t,\ \det g = u\} \#\{a \in Y,\ \operatorname{Tr} a = t,\ Na = u\}$$

$$= M \sum_{(t,u)} \frac{M^4}{M^2} h'_X(t,u) \frac{M^2}{M^2} h''_Y(t,u)$$

$$= MM^2 \frac{M^4}{M^2} \frac{M^2}{M^2} \iint_{Z(M)^2} h'_X(t,u) h''_Y(t,u)\,d\mu_{Z(M)^2}(t,u)$$

$$= M^5 \iint_{Z(M)^2} h'_X(t,u) h''_Y(t,u)\,du\,dt$$

thereby proving the desired expression for the numerator. As for the denominator, we have

$$|S(M)| = \#\{(g,a)\in X\times Y,\ \det g = Na\}$$

$$= \sum_{(t',t'',u)} \#\{g\in X,\ \det g = u,\ \text{tr } g = t'\}\ \#\{a\in Y,\ Na = u,\ \text{Tr } a = t''\}$$

$$= \sum_{(t,t,u)} \frac{M^4}{M^2} h'_X(t',u)\ \frac{M^2}{M^2} h''_Y(t'',u)$$

$$= M^5 \iiint_{Z(M)^3} h'_X(t',u)\, h''_Y(t'',u)\, du\, dt'\, dt''$$

as was to be shown.

We may then pass to the limit. We define

$$C_M(S) = \lim_{n\to\infty} C_{(M^n)}(S).$$

It is also convenient to use the abbreviations

$$\text{Num}_M(X\times_N Y) = \text{Num}_M(X,Y) = \iint_{Z_M^2} h'_X(t,u)\, h''_Y(t,u)\, dt\, du$$

$$\text{Den}_M(X\times_N Y) = \text{Den}_M(X,Y) = \iiint_{Z_M^3} h'_X(t',u)\, h''_Y(t'',u)\, du\, dt'\, dt''.$$

Suppose that S is decomposed as a disjoint union of fibered products

$$S = \bigcup_i X_i \times_N Y_i.$$

Then we define

$$\text{Num}_M(S) = \sum_i \text{Num}_M(X_i, Y_i) \quad\text{and}\quad \text{Den}_M(S) = \sum_i \text{Den}_M(X_i, Y_i).$$

At the end of the next section, we shall prove:

Theorem 6.2. *Assume that the subset S of $M_M \times_N o_M$ is stable and decomposed at level M, as a disjoint union*

$$S = \bigcup X_i \times_N Y_i \; .$$

Then

$$C_M(S) = \text{Num}_M(S)/\text{Den}_M(S) \; .$$

In the applications, the set S will be the lifting $\tilde{\mathcal{G}}$ of the Galois group $\mathcal{G}$ in the product

$$\prod [GL_2(Z_\ell) \times_N o_\ell^*] \; .$$

We know from §3 that there always exists some integer $M \geq 3$ which splits and stabilizes $\rho = (\rho', \rho'')$. It is therefore also convenient to introduce the notation

$$\text{Num}_M = \text{Num}_M(GL_2(Z_M), o_M^*) \; ,$$

giving what we call the **generic numerator**, and similarly for the denominator.

In Theorem 4.1, we had found the expression

$$\frac{1}{M} F_M(t, t) = M \frac{|\tilde{\mathcal{G}}(M)_{t,t}|}{|\tilde{\mathcal{G}}(M)|} \; .$$

The sum of this expression over $t \mod M$ is the M-th approximation to the finite part of the constant $C^{\text{fin}}(k, \rho)$, and is precisely equal to $C_{(M)}(S)$, where $S = \tilde{\mathcal{G}}$. In the next sections we shall discuss the (obvious) multiplicativity, and specific values of the integrals giving the limit value of the constant ℓ-adically. In the light of these results, we can then state:

Theorem 6.3. *Let $M \geq 3$ split and stabilize ρ, and assume that $\tilde{\mathcal{G}}$ is decomposed at level M. Then the finite part of the constant $C^{\text{fin}}(k, \rho)$ is equal to a product*

$$C^{\text{fin}}(k, \rho) = C_M(\tilde{\mathcal{G}}) \prod_{\ell \nmid M} C_\ell(\tilde{\mathcal{G}}_\ell) \; ,$$

where for $\ell \nmid M$,

$$\tilde{\mathcal{G}}_\ell = GL_2(Z_\ell) \times_N o_\ell^* \; ,$$

and $C_\ell = \text{Num}_\ell/\text{Den}_\ell$.

We observe that for almost all ℓ the set G_ℓ consists of only one piece, which is the fiber product of the full groups of invertible elements on both sides.

Furthermore, the computation of the denominators need only be done in the "generic" case, because all cases can be reduced to this one without further computation, by means of the next theorem.

Theorem 6.4. *Let*

$$\text{Den}_M = \text{Den}_M(GL_2(Z_M) \times_N o_M^*).$$

Let e *be the index of* $\tilde{\mathcal{G}}_M$ *in* $GL_2(Z_M) \times_N o_M^*$. *Then*

$$\text{Den}_M(\tilde{\mathcal{G}}_M) = e^{-1} \text{Den}_M.$$

Proof. This is actually obvious by taking the limit from finite levels, without integration, namely

$$\text{Den}_M(\tilde{\mathcal{G}}_M) = \lim_n [M^{-5n} |\tilde{\mathcal{G}}(M^n)|].$$

The index has a simple expression in most cases.

Theorem 6.5. *If* $\tilde{\mathcal{G}}_M = G_M \times_N \tilde{\mathcal{Q}}_M$, *then*

$$(GL_2(Z_M) \times_N \tilde{\mathcal{Q}}_M : \tilde{\mathcal{G}}_M) = (GL_2(Z_M) : G_M).$$

Proof. This is an immediate consequence of the standard isomorphism theorem for groups, and the fact that G_M, $GL_2(Z_M)$ have the same image under the norm map N.

The numerator will have to be computed piece by piece. In any given piece

$$X \times_N Y$$

the integral for the numerator can be written unsymmetrically by using the definition of the direct image of Haar measure,

$$\iint_{Z_M^2} h'_X(t, u) h''_Y(t, u) \, dt \, du = \int_Y h'_X(\text{Tr } z, Nz) \, dz,$$

$$= \int_X h''_Y(\text{tr } \sigma, \det \sigma) \, d\sigma,$$

where dz is Haar measure on $\mathfrak{o}_M$ and $d\sigma$ is Haar measure on M_M. We call this the seesaw principle. *For computations, only the first integral, taken over Y, will play a role.* We then write h instead of h′ to simplify the notation. From the positive values of h'_X near the identity, it will be obvious that the numerator is not equal to 0, and hence that the constant is not 0. Since the integral for the numerator has an obvious independent interest from the context in which it occurs here, we suggest the general notation

$$H(X, Y) = \int_Y h_X(\text{Tr } z, Nz) \, dz \ .$$

Fix a prime ℓ, and consider the case $M = \ell$. The map

$$T' : \text{Mat}_2(\mathbf{Q}_\ell) \longrightarrow \mathbf{Q}_\ell^2$$

is such that two non-scalar matrices have the same image under this map if and only if they are conjugate. Thus T′ parametrizes conjugacy classes over $GL_2(\mathbf{Q}_\ell)$. We define the rational Cartan subset $\mathcal{C}(k_\ell)$ determined by k to be the set of semisimple matrices $\sigma \in GL_2(\mathbf{Q}_\ell)$ such that

$$(\text{tr } \sigma, \det \sigma) = (\text{Tr } a, Na)$$

for some $a \in k_\ell^*$. If we omitted the condition of semisimplicity, we would obtain a larger set, containing certain unipotent elements, but which differs from the Cartan set by a set of measure 0.

The Cartan subset defined above is the rational one, and is invariant under conjugation by $GL_2(\mathbf{Q}_\ell)$. In our problem, we deal with its integral points, namely the set

$$\mathcal{C} = \mathcal{C}_\ell = \mathcal{C}(k_\ell) \cap GL_2(\mathbf{Z}_\ell) \ ,$$

which will be called the Cartan subset of $GL_2(\mathbf{Z}_\ell)$ determined by k, or simply the Cartan subset.

In case

$$\tilde{\mathcal{G}}_\ell = GL_2(\mathbf{Z}_\ell) \times_N \mathfrak{o}_\ell^*$$

we can write the numerator in the form

$$\text{Num}_\ell(\tilde{\mathcal{G}}_\ell) = \int_{\mathcal{C}_\ell} h''(\text{tr } \sigma, \det \sigma) \, d\sigma \ .$$

Following a suggestion of Langlands, we show that the function h'_M is the Harish transform with respect to the Cartan subgroup of the characteristic function of the set $\mathcal{C}_\ell$. As this is not used in the sequel, and is included only for the convenience of the reader who wants to connect with the literature on representation theory, we do not bother to normalize Haar measure on coset spaces carefully, and the following equalities between integrals are meant to hold only up to such normalizing constant factors.

For the rest of this section, we change notation to conform a little more to the formalism of Lie groups. So we use G to denote $GL_2(\mathbb{Q}_\ell)$. We let B be the Cartan subgroup of G corresponding to k. The Harish transform is defined to be

$$H^B \psi(b) = |D(b)|^{\frac{1}{2}} \int_{B \backslash G} \psi(g^{-1} b g) \, d\dot{g},$$

where $d\dot{g}$ is Haar measure on $B \backslash G$, normalized so that

$$dg = d\dot{g} \, db.$$

An integral formula computing differentials shows that for any function ψ,

$$\int_{B^G} \psi(\sigma) \, d\sigma = \int_B H^B \psi(b) |D(b)|^{\frac{1}{2}} \, db.$$

Warning: We use $D(b)$ to denote the discriminant of b, which is the *square* of the difference of eigenvalues, to fit the notation of §8.

If f is a function invariant under conjugation, then

$$H^B(f\psi) = f H^B(\psi).$$

Replace ψ by $f\psi$, and then let ψ be the characteristic function of $\mathcal{C}_\ell$. Then we obtain

$$\int_{\mathcal{C}} f(\sigma) \, d\sigma = \int_B f(b) H^B \psi(b) |D(b)|^{\frac{1}{2}} \, db.$$

In §8, following Theorem 8.1, we shall see that

$$|D(z)|^{\frac{1}{2}} \, dz = dt \, ds.$$

Hence we obtain

$$\int_{\mathcal{C}} f_*(\mathrm{tr}\,\sigma, \det \sigma)\, d\sigma = \iint f_*(t,s)(H^B\psi)_*(t,s)\, dt\, ds \, .$$

The lower star indicates the same function with respect to the change of variables
By definition of the direct image of Haar measure, the left hand side is also equal to

$$\iint_{T'(\mathcal{C})} f_*(t,s)\, h'(t,s)\, dt\, ds \, .$$

This proves that

$$H^B\psi = h' \, ,$$

as we wanted.

The Harish transform of certain functions on SL_2 in the ℓ-adic case has been computed before, see Sally and Shalika [Sa-Sh 2], and also [Sa-Sh 1], [Sa-Sh 3]. In a sense, the next two sections possibly perform equivalent computations, but it would not have helped to refer to the literature at this point. Furthermore, we need more complete and systematic values for the density function h'_X, with small sets X, than would in any case be available, and we need them in a form which makes the connection with the quadratic field involved easily apparent.

The Harish transform in the higher dimensional case is not yet completely cleared up, cf. Harish-Chandra [H-C].

COMPUTATIONS OF HARISH TRANSFORMS

§7. **Haar measure under the trace-determinant map on** Mat_2. **General formalism.**

Let M denote Mat_2. We abbreviate

$$M_\ell = \text{Mat}_2(Z_\ell) \quad \text{and} \quad M_M = \prod_{\ell \mid M} M_\ell \,.$$

This section deals only with matrices, so we let

$$T = (\text{tr}, \det)$$

be the map sending a matrix to its trace and determinant. Then

$$T : M_M \longrightarrow Z_M^2 \,.$$

Let R be an open subset of M_M. Then $R(M) \subset M(M)$ is the reduction of R mod M.

Given $(t,s) \in Z_M^2$ we let $R_{t,s}$ = subset of elements $\sigma \in R$ such that

$$\text{tr}\,\sigma = t \quad \text{and} \quad \det \sigma = s \,.$$

Haar measure on any compact group will be assumed to be normalized to give the group measure 1, unless otherwise specified. Let μ be Haar measure on M_M. We let μ_R be the restriction of μ to R, and 0 outside R. We shall see that the direct image $T_*\mu$ is represented by a continuous function. We define: h_R = density function of $T_*\mu_R$ with respect to Haar measure on Z_M^2. (We write h_R instead of h'_R in this section.)

Lemma 1. *Suppose that* $M = M_1 M_2$ *where* $(M_1, M_2) = 1$, *and that* $R = R_1 \times R_2$ *is a direct product of open sets in* M_{M_1} *and* M_{M_2} *respectively. Then*

$$h_R = h_{R_1} \otimes h_{R_2} \,.$$

Proof. Obvious.

Scalar matrices will play a special role in determining the density function. For any scalar a, we let

$$\psi_a : (t, s) \longmapsto (t - 2a, a^2 - at + s).$$

This mapping ψ_a describes how the trace and determinant change under translation by a scalar matrix, and has ψ_{-a} for its inverse. We note that ψ_a preserves additive Haar measure on Z_ℓ^2.

Lemma 2. (i) *Let* $\tau \in \prod_{\ell | M} GL_2(Z_\ell)$. *Then*

$$h_{\tau R \tau^{-1}} = h_R.$$

(ii) *Let* $a \in \prod_{\ell | M} Z_\ell$. *Then*

$$h_{aI+R}(t, s) = h_R(t - 2a, a^2 - at + s) = h_R(\psi_a(t, s)).$$

Proof. Again obvious.

The prime power case

We now assume that $M = M_\ell$, where ℓ is a prime. We shall compute h_R locally, whence globally. We shall see that h depends on how close an element is to being a scalar matrix. Thus it is natural to consider a filtration of matrices according to their distance from scalar matrices.

We first compute h_R when R is the inverse image under reduction mod ℓ^n of some non-scalar matrix. By translation and dilation we can get h_R when R is the set of matrices congruent to a scalar modulo a high power of ℓ. The values are given in Lemmas 5 and 7.

In the next section, we then combine these computations with the quadratic field correspondence, i.e. give the values for

$$h_R(Tr\ z, Nz)$$

when $z \in o_\ell^*$.

All the way through, the number $1/\ell$ occurs to various powers, so it is convenient to use a special notation for it, and we put

$$r = r(\ell) = 1/\ell.$$

We denote reduction mod ℓ by a bar:

$$\sigma \longmapsto \bar{\sigma}, \qquad t \longmapsto \bar{t}.$$

Theorem 7.1. *Let* $\sigma \in M$ *be such that* $\bar{\sigma}$ *is not scalar. Let* $R = \sigma + \ell^n M$, *for some integer* $n \geq 1$. *Then*

$$h_R(t,s) = \begin{cases} \ell^{2n} & \text{if } (t,s) \in T(R), \text{ or equivalently,} \\ & \text{if } (t,s) \equiv T(\sigma) \bmod \ell^n M \\ 0 & \text{otherwise}. \end{cases}$$

Proof. After a conjugation, using Lemma 2, we may assume that

$$\bar{\sigma} = \begin{pmatrix} \bar{a} & \bar{b} \\ \bar{c} & \bar{d} \end{pmatrix} \qquad \text{with} \qquad \bar{b} \neq 0.$$

We select

$$\sigma_0 \in R, \qquad \sigma_0 = \begin{pmatrix} a & b \\ c & d \end{pmatrix}$$

where b is a unit. Let $(t_0, s_0) = T(\sigma_0)$. Define the coordinate maps

$$f : Z_\ell^4 \longrightarrow R \qquad \text{by} \qquad f(w,x,y,z) = \sigma_0 + \ell^n \begin{pmatrix} w & x \\ y & z \end{pmatrix}$$

and

$$g : Z_\ell^2 \longrightarrow Z_\ell^2 \qquad \text{by} \qquad g(u,v) = (t_0 + \ell^n u, s_0 + \ell^n v).$$

Then

$$T = g T_0 f^{-1} \qquad \text{where} \qquad T_0 : Z_\ell^4 \longrightarrow Z_\ell^2$$

is given by

$$T_0(w,x,y,z) = (w+z, az+wd-cx-by + \ell^n(wz-xy))$$
$$= (w+z, wd-cx + (a+\ell^n w)z - (b+\ell^n x)y).$$

We have the commutative diagram:

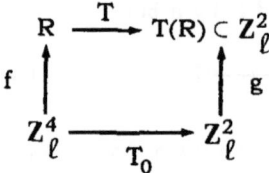

Define

$$\phi : Z_\ell^4 \longrightarrow Z_\ell^4$$

by

$$\phi(w, x, y, z) = (w, x, w+z, az + wd - cx - by + \ell^n(wz - xy))$$
$$= (w, x, T_0(w, x, y, z)) .$$

Because $b + \ell^n x$ is a unit, we can write a formula for ϕ^{-1} by solving for z and y, so ϕ is a bijection. Its Jacobian satisfies

$$|\text{Jac}_\phi| = |b + \ell^n x| = 1 ,$$

where the absolute value is ℓ-adic. Consequently ϕ is measure preserving. Then

$$T_0 = \text{pr}_{3,4} \circ \phi .$$

But f, g have constant Jacobian determinant, which is an obvious power of ℓ, and yields the value asserted in the theorem.

Let S_0 be the set of matrices $\sigma \in M$ such that $\bar{\sigma}$ is not scalar. Then S_0 is a union of sets $\sigma_i + \ell M$, of type considered in the previous theorem with $n = 1$. It is standard linear algebra that two non-scalar matrices over a field have the same characteristic polynomial if and only if they are conjugate by an element of GL_2 over that field. Over the prime field, one can count these matrices. Let $n(t, s)$ be the number of matrices σ in $M(\ell)$ such that

$$T(\bar{\sigma}) = (\bar{t}, \bar{s}) .$$

Then $n(t, s)$ is the index of the centralizer of $\bar{\sigma}$ in $GL_2(F_\ell)$. The centralizer is easily determined according to three types for σ (diagonal, unipotent, non-split Cartan), and $n(t, s)$ depends on the behavior of the polynomial

$$X^2 - \bar{t}X + \bar{s} \quad \text{over} \quad F_\ell .$$

One finds:

Lemma 3.
$$n(t, s) = \begin{cases} \ell^2 + \ell & \text{if two roots in } F_\ell \\ \ell^2 - 1 & \text{if one root in } F_\ell \\ \ell^2 - \ell & \text{if zero roots in } F_\ell. \end{cases}$$

Theorem 7.2.
$$h_{S_0}(t, s) = \begin{cases} 1 + r & \text{if } X^2 - \bar{t}X + \bar{s} \text{ has two roots in } F_\ell \\ 1 - r^2 & \text{if one root} \\ 1 - r & \text{if zero roots}. \end{cases}$$

Proof. We sum the constant value found in Theorem 7.1 over all the non-scalar matrices of $M(\ell)$, and see that the desired answer is the product of ℓ^{-2} and $n(t, s)$, found above.

The function h_{S_0} depends only on the discriminant

$$\Delta(t, s) = t^2 - 4s,$$

and in fact on the discriminant reduced mod ℓ. We shall write

$$\boxed{h_{S_0}(t, s) = h_0(\Delta(t, s)).}$$

We observe that this discriminant is invariant under "translations," precisely,

$$\Delta(t, s) = \Delta(\psi_a(t, s)).$$

We can then give the density function for the translation of certain sets by scalar matrices by using this discriminant.

The next lemma shows what happens under dilation.

Lemma 4. Let R be an open subset of M. Then

$$h_{\ell^j R}(t, s) = \begin{cases} \ell^{-j} h_R(\ell^{-j} t, \ell^{-2j} s) & \text{if } \ell^j | t \text{ and } \ell^{2j} | s \\ 0 & \text{otherwise}. \end{cases}$$

Proof. We can factor T on R as in the following diagram:

$$\begin{array}{ccc} R & \xrightarrow{\ell^j} & \ell^j R \\ T \downarrow & & \downarrow T \\ Z_\ell \times Z_\ell & \longleftarrow & \ell^j Z_\ell \times \ell^{2j} Z_\ell \end{array}$$

where the bottom map is (ℓ^{-j}, ℓ^{-2j}). The absolute value of the Jacobian of the top map is ℓ^{-4j}, and the absolute value of the Jacobian of the bottom map is ℓ^{3j}, so the lemma is clear.

Writing the expression of Lemma 4 in terms of the discriminant, and making a translation by a scalar matrix aI, we obtain:

Lemma 5. *Let* $X = aI + \ell^j S_0$, *for some integer* $j \geq 0$. *Then*

$$h_X(t,s) = \begin{cases} \ell^{-j} h_0(\ell^{-2j} \Delta(t,s)) & \text{if } (\ell^j, \ell^{2j}) \text{ divides } \psi_a(t,s) \\ 0 & \text{otherwise}. \end{cases}$$

The value h_X found in Lemma 5 is the pivotal value, which gives the density function locally.

We define S_j = set of matrices $\sigma \in M$ such that

$$\sigma \equiv \text{scalar matrix} \mod \ell^j \text{ but not} \mod \ell^{j+1}.$$

Then we can write S_j as a disjoint union

$$S_j = \bigcup_{a \bmod \ell^j} aI + \ell^j S_0.$$

Lemma 6. *If* $(t,s) \in T(S_j)$, *then there exists a unique* $a \mod \ell^j$ *such that*

$$(t,s) \in T(aI + \ell^j S_0).$$

Proof. The assertion is obvious if ℓ is odd, and in this case we need only the trace to characterize a. In fact, we get

$$a = t/2.$$

In general, suppose we have $\sigma = aI + \ell^j \sigma_0$ and $\sigma' = a'I + \ell^j \sigma'_0$ such that

$$(t, s) = T(\sigma) = T(\sigma').$$

Then

$$\sigma - aI \equiv 0 \bmod \ell^j \quad \text{and} \quad \sigma' - aI = (a' - a)I + \ell^j \sigma'_0.$$

Then

$$\begin{aligned}
\det(\sigma - aI) &= \det \sigma - a \operatorname{tr} \sigma + a^2 \\
&= \det(\sigma' - aI) \\
&\equiv (a' - a)^2 + (a' - a)\ell^j \operatorname{tr} \sigma'_0 \pmod{\ell^{2j}} \\
&\equiv 0 \pmod{\ell^{2j}}
\end{aligned}$$

because $\sigma - aI \equiv 0 \bmod \ell^j$. Therefore

$$(a' - a)^2 \equiv (a' - a)\ell^j t'_0 \pmod{\ell^{2j}}$$

where $t'_0 = \operatorname{tr} \sigma'_0$. In any case we have $a' - a \equiv 0 \bmod \ell^{j-1}$. If this congruence cannot be improved to ℓ^j then either t'_0 is divisible by ℓ and we conclude that $(a' - a)^2 \equiv 0 \pmod{\ell^{2j}}$, whence the desired congruence between a' and a, or t'_0 is not divisible by ℓ, and the right hand side of the congruence cannot be a square mod ℓ^{2j}, whence a contradiction which proves the lemma.

From Lemma 6 and Lemma 5 we conclude at once:

Lemma 7.

$$h_{S_j}(t, s) = \begin{cases} \ell^{-j} h_0(\ell^{-2j} \Delta(t, s)) & \text{if } (t, s) \in T(S_j) \\ 0 & \text{otherwise}. \end{cases}$$

Of course there is no uniqueness of the index j such that $(t, s) \in T(S_j)$, and we have:

Lemma 8. *If $(t, s) \in T(S_m)$ and $j < m$ then $(t, s) \in T(S_j)$.*

Proof. Say $(t, s) = T(\sigma)$ where $\sigma = aI + \ell^m \sigma_0$, and write

$$\ell^{m-j} \sigma_0 = \begin{pmatrix} a & b \\ c & d \end{pmatrix}.$$

If $b \neq 0$, and $b = \ell^i u$ where u is a unit, replace $\ell^{m-j}\sigma_0$ by

$$\begin{pmatrix} a & u \\ \ell^i c & d \end{pmatrix}.$$

If $b = 0$, replace c by 1. This proves the lemma.

We note that in Lemma 7, the values for h_{S_j} approach 0 as $j \to \infty$. It is therefore a reasonable convention to let

$$S_\infty = \text{set of scalar matrices}.$$

Since S_∞ has measure 0, it follows that

$$h_{S_\infty}(t, s) = 0.$$

Observe that M is the disjoint union

$$M = \bigcup_{j=0}^{\infty} S_j,$$

and so we get

$$h_M = \sum_{j=0}^{\infty} h_{S_j}.$$

By Lemma 7 and Theorem 7.2 we know that $h_{S_j} \leq 2r^j$, and therefore the series converges uniformly, defining h_M as a continuous function. It is also clear that h_M is locally constant at any point (t, s) such that $\Delta(t, s) \neq 0$, because in this case, (t, s) lies in $T(S_j)$ for only a finite number of j. On the other hand, if $\Delta(t, s) = 0$, it is easy to see that $(t, s) \in T(S_j)$ for all j, so h_M may not be locally constant at (t, s).

The preceding computations give us the values of h_X for the calculations we want to make in specific cases. They are also sufficient to prove further lemmas concerning this density function, describing how it behaves, and necessary to prove the limit theorem 6.2. However, the reader who wishes to omit the proof of Theorem 6.2 may omit the rest of this section without impeding the understanding of the rest of the paper.

Let us say that a subset of M is elementary if it is of the form

$$\sigma + \ell^n M$$

for some integer $n \geq 1$. If R is such an elementary set, and does not contain a scalar matrix, then σ is not a scalar. We may then write σ in the form

$$\sigma = aI + \ell^j \sigma_0 + \ell^n M ,$$

where σ_0 is not scalar mod ℓ, and it follows that $j < n$.

On the other hand, if R contains a scalar, then R is of the form

$$R = aI + \ell^n M .$$

If X is a subset of M, we denote by $X(\ell^n)$ its reduction mod ℓ^n. We recall that X is called stable of level m if it is the inverse image of its reduction mod ℓ^m. (As we deal with the prime power case, we use the exponent to index the stability level.) We say that X is stable if it is stable at some level. A set is stable if and only if it is a finite union of elementary sets.

Theorem 7.3. *If X is a stable set in M then h_X is continuous, and locally constant at any point (t, s) which does not lie in $T(X \cap S_\infty)$. In particular, if X contains no scalar matrix, then h_X is locally constant.*

Proof. We already know the theorem when $X = M$. An elementary set is obtained from M by translations and dilations, to which we can apply Theorem 7.1 and Lemmas 2, 4, which show the theorem to be true for elementary sets. Since a stable set is a finite union of elementary sets, the theorem follows at once.

Theorem 7.4. *For any stable set X,*

$$\lim_{n \to \infty} h_{X(\ell^n)}(t, s) = h_X(t, s) .$$

Proof. Let X be stable at level m and let $n \geq m$. Let

$$E_n = (t, s) + \ell^n Z_\ell^2 .$$

Then

$$\ell^{-2n} h_{X(\ell^n)}(t, s) = \int_{E_n(\ell^n)} h_{X(\ell^n)}(t', s') dt' ds'$$

$$= \mu_n(X(\ell^n) \cap T_n^{-1}(E_n(\ell^n))) ,$$

where μ_n is Haar measure on $M(\ell^n)$, and T_n is now indexed by n for clarity, being the trace-determinant map on $M(\ell^n)$. Since X is assumed stable at level n, it follows that
$$X \cap T^{-1}(E_n)$$
is also stable at level n, and hence our last expression is equal to
$$\mu(X \cap T^{-1}(E_n)) = \int_{E_n} h_X(t', s')\, dt'\, ds',$$
where μ is Haar measure on M. Thus we have shown that
$$h_{X(\ell^n)}(t, s) = \frac{1}{\mu_n(E_n)} \int_{E_n} h_X(t', s')\, dt'\, ds'.$$
Taking the limit as $n \to \infty$ we see that the right hand side approaches $h_X(t, s)$, as was to be shown.

Theorem 7.5. *Let Y be a stable set in o_ℓ, and X stable in M. Then*
$$\lim_{n \to \infty} \int_{Y(\ell^n)} h_{X(\ell^n)}(\operatorname{Tr} z, Nz)\, dz = \int_Y h_X(\operatorname{Tr} z, Nz)\, dz.$$

Proof. The expression
$$h_{X(\ell^n)}(\operatorname{Tr} z, Nz)$$
defines a function on $o(\ell^n)$. Suppose that Y is stable at level n. Then
$$\int_{Y(\ell^n)} h_{X(\ell^n)}(\operatorname{Tr} z, Nz)\, dz = \int_Y h_{X(\ell^n)}(\operatorname{Tr} z, Nz)\, dz.$$
Theorem 7.4 and the bounded convergence theorem conclude the proof.

This is the result which we needed to prove Theorem 6.2, expressing the basic constant as a limit from finite levels. Note that the handling of the denominator is trivial, by using Theorem 6.4, and the fact that the generic Galois group is stable at level 1.

§8. Relations with the trace-norm map on k

We have $o_\ell = Z_\ell \otimes o$. We write $o_\ell = [1, \eta]$ to mean that 1 and η form a basis for o_ℓ over Z_ℓ. We can find such a basis with $\eta \in o$. We let

$$T'' = (\text{Tr}, N)$$

be the trace-norm map from k to Q. Then T'' extends uniquely to o_ℓ, giving rise to a map still denoted by T'', namely

$$T'' : o_\ell \longrightarrow Z_\ell^2 .$$

An element $z \in o_\ell$ is then a root of its characteristic equation

$$X^2 - (\text{Tr } z)X + Nz = 0 .$$

The automorphism of k extends by continuity to k_ℓ, and induces an automorphism of o_ℓ. We can define the discriminant

$$D(z) = D_\ell(z) = (z - \bar{z})^2 .$$

Then

$$D(z) = \Delta(\text{Tr } z, Nz) = (\text{Tr } z)^2 - 4Nz .$$

Let $L_z = [1, z]$ be the sublattice (over Z_ℓ) generated by 1 and z. Then its discriminant is well defined modulo the square of a unit in Z_ℓ, and we have

$$\text{Discr}(L_z) = D(z)(\text{unit})^2 .$$

There is an integer $j \geq 0$ such that

$$D(z) = \ell^{2j}(\text{unit})^2 D ,$$

where the unit is of course an ℓ-adic unit, i.e. an element of Z_ℓ^*, and $D = D(o)$ is the discriminant of k. We define

$$v(z) = j$$

to be this integer.

Lemma 1. *Let $o_\ell = [1, \eta]$ and $z = x + y\eta$, where $x, y \in Z_\ell$. The following conditions are equivalent:*

(i) $v(z) = j$.
(ii) $y = \ell^j u$ *for some unit* $u \in Z_\ell^*$.
(iii) $z \equiv a \pmod{\ell^j o_\ell}$ *for some* $a \in Z_\ell$, *but z is not congruent to any element of* Z_ℓ *mod* ℓ^{j+1}.

Proof. This is a simple exercise, which will be left to the reader.

In classical terminology, we could call ℓ^j the **conductor** of z, or the conductor of L_z. Then $v(z)$ is the order of the conductor.

Observe that if we coordinatize z by (x, y) such that

$$z = x + y\eta, \qquad x, y \in Z_\ell$$

and put $r = 1/\ell$, then

$$|y| = r^{v(z)}.$$

This (ℓ-adic) absolute value is independent of the choice of basis $o_\ell = [1, \eta]$. With such coordinates, we have

$$\boxed{dz = dx\, dy\,.}$$

Lemma 2. *Let $z \in o_\ell$. Then*

$$(\operatorname{Tr} z, Nz) \in T'(S_j) \quad \text{if and only if} \quad j \leq v(z).$$

Proof. Let

$$(\operatorname{Tr} z, Nz) = (\operatorname{tr} \sigma, \det \sigma)$$

with some matrix $\sigma = aI + \ell^j \sigma_0$, and σ_0 not scalar mod ℓ. Let

$$w = \ell^{-j}(z - a).$$

Then $(\operatorname{Tr} w, Nw) = (\operatorname{tr} \sigma_0, \det \sigma_0)$. Hence $w \in o_\ell$, and

$$z = a + \ell^j w,$$

so that $v(z) \geq j$ as desired.

Conversely, let $v = v(z)$. We can write

$$z = a + \ell^v w, \qquad \text{with} \qquad a \in Z_\ell, \ w \in o_\ell$$

and $w \in Z_\ell$. Under the regular representation on $o_\ell = [1, \eta]$, we obtain representing matrices such that

$$\sigma_z = aI + \ell^v \sigma_w,$$

which show that $(\operatorname{Tr} z, Nz) \in T'(S_m)$ for $m \geq v$. The first part of the proof shows that $m = v$. If $j \leq v$ we can use Lemma 8 of the preceding section to conclude the proof.

From Lemma 7 of the preceding section, we can write h_{S_j} in the form

$$h_{S_j}(\operatorname{Tr} z, Nz) = \begin{cases} \ell^{-j} h_0(\ell^{-2j} D(z)) & \text{if } j \leq v(z) \\ 0 & \text{otherwise.} \end{cases}$$

Hence we get h_{S_j} in a form convenient for computations:

Lemma 3. *Let* $r = r(\ell) = 1/\ell$.

$$h_{S_j}(\operatorname{Tr} z, Nz) = \begin{cases} r^j(1-r^2) & \text{if } j < v(z) \\ 0 & \text{if } j > v(z) \end{cases}$$

and for $j = v(z) = v$:

$$h_{S_v}(\operatorname{Tr} z, Nz) = \begin{cases} r^v(1+r) & \text{if } \left(\frac{k}{\ell}\right) = 1 \\ r^v(1-r^2) & \text{if } \left(\frac{k}{\ell}\right) = 0 \\ r^v(1-r) & \text{if } \left(\frac{k}{\ell}\right) = -1. \end{cases}$$

Taking into account Lemma 8 of the preceding section, and summing a geometric series gives the value of h_M for the full set of matrices.

Let $M = M_\ell$ and $v = v(z)$. Then

$$h_M(\operatorname{Tr} z, Nz) = (1+r)(1-r^v) + h_{S_v}(\operatorname{Tr} z, Nz).$$

It is sometimes convenient to combine the terms involving $r^{v(z)}$ when we integrate. Thus we define the function

$$\psi_\varrho(k) = \begin{cases} 0 & \text{if } \left(\frac{k}{\ell}\right) = 1 \\ r(1+r) & \text{if } \left(\frac{k}{\ell}\right) = 0 \\ 2r & \text{if } \left(\frac{k}{\ell}\right) = -1 \end{cases}$$

and get:

Theorem 8.1.

$$\boxed{h_M(\operatorname{Tr} z, Nz) = 1 + r - \psi_\varrho(k)\, r^{v(z)}}$$

Remark. Although we do not need it for the sequel, it may be convenient for the reader to see also the formula for h'', which reads:

$$h''(\operatorname{Tr} z, Nz) = 2|D|^{-1}\, r^{-v(z)}.$$

The proof is easy. The Jacobian determinant of T'' is immediately computed to be

$$4yN\eta - y(\operatorname{Tr} \eta)^2 = -yD.$$

Its absolute value is therefore $|D|\,|y|$. Since T'' is two-to-one, the formula follows at once. As $v(z) \to \infty$, note that

$$h(\operatorname{Tr} z, Nz) = h'(\operatorname{Tr} z, Nz)$$

remains bounded, and converges, while $h''(\operatorname{Tr} z, Nz)$ tends to infinity. The way we choose our coordinates, and write the integral for the numerator $\operatorname{Num}_\varrho(X, Y)$, the density functions will always appear in their bounded form, i.e. $h'(\operatorname{Tr} z, Nz)\,dz$

We have the corresponding lemmas giving the behavior under dilation and translation, easier to handle when dealing with the quadratic field than with matrices.

Lemma 4. *Let R be an open set in M. Then*

$$h_{\ell^j R}(\operatorname{Tr} z, Nz) = \begin{cases} \ell^{-j} h_R(\operatorname{Tr}(z/\ell^j), N(z/\ell^j)) & \text{if } z \in \ell^j o_\varrho \\ 0 & \text{otherwise.} \end{cases}$$

If Y is open in o_ℓ, then

$$\int_Y h_{\ell^j R}(\mathrm{Tr}\, z, Nz)\, dz = r^{3j} \int_{\ell^{-j}Y \cap o_\ell} h_R(\mathrm{Tr}\, z, Nz)\, dz .$$

Proof. The formula for $h_{\ell^j R}$ is obvious from Lemma 4 of the preceding section. We then make a change of variables by a dilation, which gives the value for the integral.

Lemma 5. *Let R be an open set in M. Let* $a \in Z_\ell$. *Then*

$$h_{aI+R}(\mathrm{Tr}\, z, Nz) = h_R(\mathrm{Tr}(z-a), N(z-a)) .$$

If Y is open in o_ℓ, *then*

$$\int_Y h_{aI+R}(\mathrm{Tr}\, z, Nz)\, dz = \int_{Y-a} h_R(\mathrm{Tr}\, z, Nz)\, dz .$$

Proof. The formula for h_{aI+R} is obvious from Lemma 2 of the preceding section. We then make a translation of variable to get the other form for the integral.

Using the two lemmas, one can always reduce an integral of the type we shall consider to Theorem 7.1 or Theorem 8.1.

Indeed, the sets X in M which we consider are finite unions of sets of type

$$X = aI + \ell^j \sigma + \ell^n M, \quad \text{or} \quad X = aI + \ell^n M,$$

where σ is not scalar mod ℓ, and $j < n$. Using Lemmas 4 and 5, we see that

$$\int_Y h_X(\mathrm{Tr}\, z, Nz)\, dz = \int_{Y'} h_{X'}(\mathrm{Tr}\, z, Nz)\, dz \quad (\text{times } r^{3j} \text{ or } r^{3n}),$$

where Y' is a suitable transform of Y, and X' has the form

$$X' = \sigma + \ell^{n-j} M \quad \text{or} \quad X' = M .$$

We carry out a simple case as an example.

Lemma 6. *Let $a \in Z_\ell$. Then*

$$\int_{o_\ell} h_{aI+\ell^n M}(\text{Tr } z, Nz)\, dz = r^{3n}\left[(1+r) - \psi_\ell(k)\frac{1}{1+r}\right].$$

Proof.

(1) $\quad\displaystyle\int_{o_\ell} h_{aI+\ell^n M}(\text{Tr } z, Nz)\, dz = \int_{o_\ell} h_{\ell^n M}(\text{Tr } z, Nz)\, dz$

(2) $\quad\displaystyle = \int_{\ell^n o_\ell} h_{\ell^n M}(\text{Tr } z, Nz)\, dz$

(3) $\quad\displaystyle = r^{3n}\int_{o_\ell} h_M(\text{Tr } z, Nz)\, dz$

(4) $\quad\displaystyle = r^{3n}\left[(1+r) - \psi_\ell(k)\frac{1}{1+r}\right]$

as shown in the next lemma.

Lemma 7. *We have*

$$\int_{o_\ell} r^{v(z)}\, dz = \frac{1}{1+r}$$

and

$$\int_{o_\ell} h_M(\text{Tr } z, Nz)\, dz = (1+r) - \psi_\ell(k)\frac{1}{1+r}.$$

Proof. The first integral is equal to

$$\iint |y|\, dx\, dy$$

taken over $Z_\ell \times Z_\ell$. This makes it obvious. The second is obtained by substituting in Theorem 8.1.

In the next theorems we give explicitly some integrals which are used to compute the constants.

Theorem 8.2. *We have the following integral table.* $(r = 1/\ell.)$

	$\int_{o_\ell^*} dz = \mu(o_\ell^*)$	$\int_{o_\ell^*} r^{v(z)} dz$
$\left(\frac{k}{\ell}\right) = 1$	$(1-r)^2$	$\frac{(1-r)(1-r-r^2)}{1+r}$
$\left(\frac{k}{\ell}\right) = 0$	$1-r$	$\frac{1-r}{1+r}$
$\left(\frac{k}{\ell}\right) = -1$	$1-r^2$	$\frac{1-r^3}{1+r}$

Proof. We have
$$\mu(o_\ell^*) = \frac{|o(\ell)^*|}{|o(\ell)|},$$
which is computed explicitly and easily in the three cases. This gives the first column. For the second column, the integral over the units is expressed as the difference of the integral over o_ℓ and the integral over the non-units. One must characterize the units in each case.

If $\left(\frac{k}{\ell}\right) = 1$, then the non-units are of the form
$$\ell Z_\ell \times Z_\ell \quad \cup \quad Z_\ell^* \times \ell Z_\ell,$$
and the union is disjoint. The integral is then trivially computed.

If $\left(\frac{k}{\ell}\right) = -1$, then the non-units are the elements such that ℓ divides both x and y, which again makes the integral obvious.

Suppose that $\left(\frac{k}{\ell}\right) = 0$, so ℓ ramifies in k. We can pick an ℓ-adic basis $[1, \lambda]$ for o_ℓ over Z_ℓ such that λ is not a unit, e.g. a local parameter at a prime above ℓ. Then the units $x + y\lambda$ with $x, y \in Z_\ell$ are precisely those elements for which x is a unit. From this the table entry for the ramified case follows at once.

The table of Theorem 8.2 suffices for situations of general type. It is inadequate for the computations of Part III, where we study specific instances needing a slight generalization. The reader is advised to skip the next theorem until he needs it.

Theorem 8.3. *Let* $a \in o_\ell$. *Then*

$$\int_{a+\ell o} r^{v(z)} dz = \begin{cases} \dfrac{r^3}{1+r} & \text{if } \bar{a} \in Z(\ell) \\ r^2 = \mu(a+\ell o) & \text{otherwise .} \end{cases}$$

Proof. Note that $x + y\eta$ lies in $Z(\ell)$ if and only if y is not a unit. In this case, the desired integral is equal to

$$\iint_{\substack{|y|<1 \\ x \equiv a}} |y|\, dx\, dy .$$

The single integral over $x \equiv a$ splits off to give the value r. For the integral over y, we write $y = \ell y'$, $dy = r dy'$, and the desired value drops out. On the other hand, if y is a unit, then we have to compute $\mu(a + \ell o)$, which is trivial.

§9. Computation of C_ℓ for almost all ℓ

We have seen that for almost all ℓ, the ℓ-adic part of the desired constant is given by a ratio of integrals

$$C_\ell = \text{Num}_\ell / \text{Den}_\ell ,$$

where the numerator is

$$\text{Num}_\ell = \int_{o_\ell^*} h'_M(\text{Tr } z, Nz)\, dz$$

and the denominator need not be handled by means of its integral expression, but from the elementary definition

$$\text{Den}_\ell = \lim_{n \to \infty} \ell^{-5n} |GL_2(\ell^n) \times_N o(\ell^n)^*| .$$

Theorem 9.1. *Let* $r = r(\ell) = 1/\ell$. *We have:*

$$\text{Den}_\ell = (1-r^2)\mu(o_\ell^*) .$$

The constant C_ℓ *and numerator* Num_ℓ *are given in the following table.*

	C_ℓ	Num_ℓ	Den_ℓ
$\left(\frac{k}{\ell}\right) = 1$	$\dfrac{1}{1-r}$	$(1+r)(1-r)^2$	$(1+r)(1-r)^3$
$\left(\frac{k}{\ell}\right) = 0$	$\dfrac{1}{1-r^2}$	$1-r$	$(1+r)(1-r)^2$
$\left(\frac{k}{\ell}\right) = -1$	$\dfrac{1}{1+r}\left[1 + \dfrac{2r^2}{(1+r)(1-r^2)}\right]$	$\dfrac{1-r}{1+r}[1+r+r^2-r^3]$	$(1+r)^2(1-r)^2$

Proof. We start with the denominator, and in this case, there is no need for the integral expression, the matter is totally elementary. We see immediately that the expression for the denominator is stable at level 1, i.e. independent of n for $n \geq 1$. Consequently

$$\text{Den}_\ell = \ell^{-5} |GL_2(Z(\ell)) \times_N o(\ell)^*| .$$

From the exact sequence

$$0 \longrightarrow GL_2(Z(\ell)) \times_N o(\ell)^* \longrightarrow GL_2(Z(\ell)) \times o(\ell)^* \longrightarrow Z(\ell)^* \longrightarrow 0$$

we get

$$\text{Den}_\ell = \frac{|GL_2(Z(\ell)) \times o(\ell)^*|}{(\ell-1)\ell^5} .$$

But

$$\mu(o^*_\ell) = \frac{|o(\ell)^*|}{|o(\ell)|} \quad \text{and} \quad |GL_2(Z(\ell))| = \ell(\ell-1)(\ell^2-1) .$$

The desired value for the denominator drops out.

Next we deal with the numerator. We have

$$\text{Num}_\ell = \int_{o^*_\ell} h_M(\text{Tr } z, Nz) dz .$$

The appropriate expression for h_M was given in Theorem 8.1, from which we see that the integral for the numerator is a sum of integrals which have been computed in Theorem 8.2. This gives the asserted value for the numerator.

§10. The constant for Serre curves, $K \cap k_{ab} = Q_{ab}$

In Part I we had already discussed the special case of Serre curves. They have a special Galois group described as follows. Let q be an odd prime. We let:

$$E_2 = \{\sigma \in GL_2(Z_2), \bar{\sigma} \text{ is an even permutation in } GL_2(2)\}$$

$$E_q = \{\sigma \in GL_2(Z_q), \det \bar{\sigma} \text{ is a square in } F_q^*\}.$$

The notation $\bar{\sigma}$ denotes the reduction of σ mod 2 and mod q respectively. We note that E_2 and E_q are of index 2 in their respective GL_2, and we let O_2 and O_q be their respective cosets, which we call the odd cosets, as distinguished from E_2 and E_q which are called the even cosets.

We let
$$S_{2q} = (E_2 \times E_q) \cup (O_2 \times O_q).$$

Serre's subgroup is by definition

$$G = S_{2q} \times \prod_{\ell \neq 2, q} GL_2(Z_\ell).$$

In this section, we want to compute the value $C_{2q}(\mathcal{S})$ as given in Theorem 6.2, where $\mathcal{S}$ is the set
$$\mathcal{S}_{2q} = \mathcal{S} = S_{2q} \times_N o_{2q}^*,$$

and $o_{2q}^* = o_2^* \times o_q^*$. The relevance of this computation for the applications was shown in Theorem 3.3.

The constant is a quotient,

$$C_{2q}(\mathcal{S}) = \text{Num}_{2q}(\mathcal{S})/\text{Den}_{2q}(\mathcal{S}).$$

The denominator is easily taken care of, by reduction to the "generic" case, handled in the preceding section.

Theorem 10.1. $\text{Den}_{2q}(\mathcal{S}) = \frac{1}{2} \text{Den}_{2q}(GL_2(Z_{2q}) \times_N o_{2q}^*)$

$$= \frac{1}{2} \text{Den}_2 \cdot \text{Den}_q,$$

where Den_2 and Den_q are the denominators computed in Theorem 9.1.

Proof. The subgroup $\mathcal{S}$ is of index 2 in the full fiber product

$$GL_2(Z_{2q}) \times_N \mathfrak{o}_{2q}^* .$$

We can then apply Theorem 6.4.

Next we deal with the numerator,

$$\mathrm{Num}_{2q}(\mathcal{S}) = \int h_{S_{2q}}(\mathrm{Tr}\ z, Nz)\, dz ,$$

and the integral is taken over $\mathfrak{o}_{2q}^2$. Since S_{2q} is a disjoint union of

$$E_{2q} = E_2 \times E_q \qquad \text{and} \qquad O_{2q} = O_2 \times O_q ,$$

we have

$$h_{S_{2q}} = h_{E_{2q}} + h_{O_{2q}} ,$$

and the integral splits as a sum of two integrals. Each of these is a product, because the variables separate, for instance

$$h_{E_2 \times E_q}(\mathrm{Tr}\ z, Nz) = h_{E_2}(\mathrm{Tr}\ z_2, Nz_2)\, h_{E_q}(\mathrm{Tr}\ z_q, Nz_q) .$$

This yields:

(1) $$\mathrm{Num}_{2q}(\mathcal{S}) = \mathrm{Num}_2(E_2 \times_N \mathfrak{o}_2^*)\, \mathrm{Num}_q(E_q \times_N \mathfrak{o}_q^*)$$
$$+ \mathrm{Num}_2(O_2 \times_N \mathfrak{o}_2^*)\, \mathrm{Num}_q(O_q \times_N \mathfrak{o}_q^*) .$$

We use abbreviations,

$$\mathrm{Num}_\ell^+ = \mathrm{Num}_\ell(E_\ell \times_N \mathfrak{o}_\ell^*) = \int_{\mathfrak{o}_\ell^*} h_{E_\ell}(\mathrm{Tr}\ z, Nz)\, dz$$

$$\mathrm{Num}_\ell^- = \mathrm{Num}_\ell(O_\ell \times_N \mathfrak{o}_\ell^*) = \int_{\mathfrak{o}_\ell^*} h_{O_\ell}(\mathrm{Tr}\ z, Nz)\, dz .$$

We also have the generic numerator,

$$\mathrm{Num}_\ell = \mathrm{Num}_\ell^+ + \mathrm{Num}_\ell^- = \mathrm{Num}_\ell(GL_2(Z_\ell) \times_N \mathfrak{o}_\ell^*) ,$$

which was computed in §9 for any prime ℓ.

With this notation, we have:

Theorem 10.2. $\text{Num}_{2q}(\mathcal{S}) = \text{Num}_2^+ \text{Num}_q^+ + \text{Num}_2^- \text{Num}_q^-$.

Furthermore, we can write

$$\text{Num}_\ell^+ = \text{Num}_\ell - \text{Num}_\ell^-,$$

so that $\text{Num}_{2q}(\mathcal{S})$ can be computed in terms of Num_ℓ, which we already know, and Num_ℓ^- (with $\ell = 2, q$) which we shall determine and tabulate. In the table, $o_{2,\text{even}}^*$ denotes the units in o_2 with even trace.

	$\mu(o_{2,\text{even}}^*)$	Num_2^-	Num_2^+
$\left(\frac{k}{2}\right) = 1$	1/4	3/16	3/16
$\left(\frac{k}{2}\right) = 0$	$\frac{1}{2}$	3/8	1/8
$\left(\frac{k}{2}\right) = -1$	1/4	3/16	17/48

The table for q follows. As usual, $r = r(q) = 1/q$.

	Num_q^-	Num_q^+
$\left(\frac{k}{q}\right) = 1$	$\frac{1}{2}(1-r)^2(1+r)$	$\frac{1}{2}(1-r)^2(1+r)$
$\left(\frac{k}{q}\right) = 0$	0	$1-r$
$\left(\frac{k}{q}\right) = -1$	$\frac{1}{2}(1-r)^2(1+r)$	$\frac{1}{2}\frac{1-r}{1+r}[1+r+3r^2-r^3]$

Computations at 2

The odd elements in $GL_2(F_2)$ are represented by the matrices

$$\begin{pmatrix} 0 & 1 \\ 1 & 0 \end{pmatrix}, \quad \begin{pmatrix} 1 & 1 \\ 0 & 1 \end{pmatrix}, \quad \begin{pmatrix} 1 & 0 \\ 1 & 1 \end{pmatrix}.$$

We see that $O_2(2)$ consists of the non-scalar matrices with even trace. Hence

$$h_{O_2}(t, u) = \begin{cases} 0 & \text{if } t \text{ is odd or } u \text{ is not a unit} \\ h_{S_0}(t, u) & \text{if } t \text{ is even and } u \text{ is a unit}. \end{cases}$$

Therefore by Theorem 7.2 we find with $r = r(2) = 1/2$:

$$h_{O_2}(t, u) = \begin{cases} 0 & \text{if } t \text{ is odd or } u \text{ is not a unit} \\ 1 - r^2 & \text{if } t \text{ is even and } u \text{ is a unit}. \end{cases}$$

Hence

(2) $$\text{Num}_2^- = \int_{o_2} h_{O_2}(\text{Tr } z, Nz) dz = (1 - r^2)\mu(o_{2,\text{even}}^*).$$

The values of $\mu(o_{2,\text{even}}^*)$ depend on $\left(\frac{k}{2}\right)$.

If 2 splits completely in k, then

$$o_2^* \approx Z_2^* \times Z_2^*$$

and every unit has even trace. Hence the measure of the units is 1/4. The rest of the top line in the table is then obtained from the value for Num_2^- in (2), and the generic numerator of Theorem 9.1.

If 2 ramifies, every unit has even trace, and the units form the single additive coset of the non-units. Hence the measure of the units is 1/2. The other table entries follow trivially.

We leave the third case to the reader.

Computations at q

By the definition of the odd elements at q, we find

$$h_{O_q}(t, u) = \begin{cases} h_{S_0}(t, u) & \text{if } \left(\frac{u}{q}\right) = -1 \\ 0 & \text{otherwise} . \end{cases}$$

If q ramifies in k, then the norm of every unit is a square, and hence we get 0 in the corresponding table entry.

The other two cases will be seen to give the same value. If $\left(\frac{u}{q}\right) = -1$ then

$$X^2 - \bar{t}X + \bar{u} = 0$$

has either zero or two roots in F_q. We have

$$\Delta(\text{Tr } z, Nz) = y^2 D \quad \text{if} \quad z = x + y\eta .$$

If Nz is a non-square unit then $\Delta(\text{Tr } z, Nz)$ is a unit because

$$(\text{Tr } z)^2 - 4Nz \not\equiv 0 \pmod{q} .$$

Hence y is a unit, and therefore

$$\left(\frac{\Delta}{q}\right) = \left(\frac{D}{q}\right) = \left(\frac{k}{q}\right) .$$

From Theorem 7.2 we therefore obtain the value:

Lemma 1.
$$h_{O_q}(\text{Tr } z, Nz) = \begin{cases} 0 & \text{unless Nz is a non-square unit} \\ & \text{and otherwise:} \\ 1 + r & \text{if } \left(\frac{k}{q}\right) = 1 \\ 1 - r & \text{if } \left(\frac{k}{q}\right) = -1 . \end{cases}$$

Let U_q^- be the subset of o_q^* consisting of those elements whose norm is not a square in Z_q^*. We find:

(3) $$\text{Num}_q^- = \iint_{o_q} h_{O_q}(\text{Tr } z, Nz) \, dz = \begin{cases} (1+r)\mu(U_q^-) & \text{if } \left(\frac{k}{q}\right) = 1 \\ (1-r)\mu(U_q^-) & \text{if } \left(\frac{k}{q}\right) = -1 . \end{cases}$$

The norm map

$$N : o_q^* \longrightarrow Z_q^*$$

is surjective. Hence

(4) $$\mu(U_q^-) = \tfrac{1}{2}\mu(o_q^*),$$

where

(5) $$\mu(o_q^*) = \begin{cases} (1-r)^2 & \text{if } \left(\tfrac{k}{q}\right) = 1 \\ 1-r^2 & \text{if } \left(\tfrac{k}{q}\right) = -1. \end{cases}$$

In view of (3), this gives the first column in the table for Num_q^-. The second column is obtained as before, by subtracting from Num_q found in Theorem 9.1.

§11. The constant for $X_0(11)$

The Galois group of division points of $X_0(11)$ was determined in Part I, §8. We recall the result, and derive implications for the correspondence with imaginary quadratic fields.

We have
$$G_\ell = GL_2(Z_\ell) \quad \text{for} \quad \ell \neq 5.$$

We have
$$G \subset G_2 \times G_5 \times G_{11} \times \prod_{\ell \neq 2,5,11} GL_2(Z_\ell),$$

and
$$G = G_{110} \times \prod_{\ell \neq 2,5,11} GL_2(Z_\ell).$$

All we have to worry about is $G_{110} \subset G_2 \times G_5 \times G_{11}$.

We denote by v_{10} the multiplicative group $Z(11)^*$. We let v_2 and v_5 be its components of orders 2 and 5 respectively.

We have a homomorphism
$$\phi_2 : G_2 \longrightarrow \{\pm 1\}$$

corresponding to the even and odd elements. We denote
$$G_{2,1} = E_2 \quad \text{and} \quad G_{2,-1} = O_2,$$

the inverse image of 1 and -1 respectively by ϕ_2 in G_2.

In Part I, §8 we had described precisely the correspondence between G_5 and G_{22} (actually factoring through G_{11}). There is a homomorphism
$$\phi_5 : G_5 \longrightarrow Z(11)^*/\pm 1 \approx v_5$$

such that, in our previous notation,
$$\phi_5(\sigma) = (\pm 2)^{\psi(\sigma)},$$

where $\psi(\sigma) = a \bmod 5$ if

$$\sigma = \begin{pmatrix} 1+5a & 5b \\ 5c & u \end{pmatrix}.$$

We let $G_{5,\zeta} = \phi_5^{-1}(\zeta)$, for $\zeta \in v_5$.

Finally, we have the natural homomorphism

$$\det{}_{(11)} = \phi_{11} : G_{11} \longrightarrow Z(11)^* \approx v_{10} = v_2 \times v_5$$

If $\zeta \in Z(11)^*$ we let $G_{11,\zeta} = \phi_{11}^{-1}(\zeta)$.

Then Theorem 8.3 of Part I can be formulated by saying that

$$\boxed{G_{110} = \bigcup_\zeta G_{11,\zeta} \times G_{2,\zeta(2)} \times G_{5,\zeta(5)}}$$

where the union is taken for $\zeta \in v_{10}$. We have used the notation

$$\zeta(2) \quad \text{and} \quad \zeta(5)$$

to denote the canonical image of ζ in v_2 and v_5 respectively. Observe that the products are automatically fibered over $Q(\sqrt{\Delta})$, which is contained in the field of 11-th roots of unity.

The next theorem gives us the denominator of the constant, by the proxies of Theorems 6.4 and 9.1.

Theorem 11.1. *For* $X_0(11)$ *we have*

$$(GL_2(Z_{110}) : G_{110}) = 1,200.$$

Proof. We note that G_{110} has index 10 in $G_2 \times G_5 \times G_{11}$. Furthermore, G_5 has index 120 in $GL_2(Z_5)$. The theorem follows at once.

We then come to the computation of the numerator.

We first treat the case when $K \cap k_{ab} = Q_{ab}$. Then all but the 110-factors split off and are generic, so what we have to compute is

$$\text{Num}(\bar{\mathcal{G}}_{110}) = \text{Num}_{110}(\tilde{\mathcal{G}}),$$

and for this we shall use a decomposition into fiber products to get:

Theorem 11.2. *Let* $\mathcal{S}_{2.11}$ *be Serre's subgroup. Assume that* $K \cap k_{ab} = Q_{ab}$. *Then*
$$\text{Num}_{110}(\tilde{\mathcal{G}}) = \frac{1}{5} \text{Num}_5(\tilde{\mathcal{G}}_5) \text{Num}_{2.11}(\mathcal{S}_{2.11}).$$

Proof. We need a lemma.

Lemma. *The values*
$$\text{Num}_5(G_{5,\zeta(5)}, \mathfrak{o}_5^*)$$
are independent of ζ, *and equal to* $\frac{1}{5} \text{Num}_5(G_5, \mathfrak{o}_5^*)$.

Proof. Let X_a be the set of matrices
$$\begin{pmatrix} 1+5a & 5b \\ 5c & u \end{pmatrix}$$
with a, b, c $\in Z_5$ and u $\in Z_5^*$. Then
$$X_a + 5dI = X_{a+d}.$$
Therefore
$$\int_{\mathfrak{o}_5^*} h_{X_{a+d}}(\text{Tr } z, Nz) dz = \int_{\mathfrak{o}_5^*} h_{5dI+X_a}(\text{Tr } z, Nz) dz$$
$$= \int_{\mathfrak{o}_5^*} h_{X_a}(\text{Tr } z, Nz) dz,$$
as was to be shown.

Returning to the theorem, we have
$$\text{Num}_{110}(\tilde{\mathcal{G}}) = \sum_\zeta \text{Num}_{11}(G_{11,\zeta}, \mathfrak{o}_{11}^*) \text{Num}_2(G_{2,\zeta(2)}, \mathfrak{o}_2^*) \text{Num}_5(G_{5,\zeta(5)}, \mathfrak{o}_5^*).$$

By the lemma, we can replace $\text{Num}_5(G_{5,\zeta(5)}, \mathfrak{o}_5^*)$ by its constant value. The sum then gives precisely the numerator for the Serre group, as was to be shown.

Theorem 10.2 gives the numerator for the Serre group, so we are reduced to computing $\text{Num}_5(\tilde{\mathcal{G}}_5)$. This is merely a matter of putting together all the techniques which we already know, but we give the details.

Theorem 11.3. *Let* $r = r(5) = 1/5$.

$$\mathrm{Num}_5(G_5, o_5^*) = \begin{cases} 7r^4 + r^3 & \text{if } \left(\frac{k}{5}\right) = 1 \\ r^3 & \text{if } \left(\frac{k}{5}\right) = 0 \\ r^3(1+r) - \frac{2r^4}{1+r} & \text{if } \left(\frac{k}{5}\right) = -1 \end{cases}$$

Proof. Let $d = 1, 2, 3, 4$. Let

$$\sigma(d) = \begin{pmatrix} 1 & 0 \\ 0 & d \end{pmatrix}.$$

Let $R(d) = \sigma(d) + 5M_5$. Then we have a disjoint union

$$G_5 = \bigcup_{d=1}^{4} R(d).$$

Define

$$f(d) = \int_{o_5} h_{R(d)}(\mathrm{Tr}\ z, Nz)\,dz.$$

Then

$$\mathrm{Num}_5(\tilde{\mathcal{G}}_5) = \sum_{d=1}^{4} f(d).$$

Let $T_{d+1,d} = \{z \in o_5,\ \mathrm{Tr}\ z \equiv d+1\ \text{and}\ Nz \equiv d \bmod 5\}$. It is immediately verified that:

If $\left(\frac{k}{5}\right) \neq 1$ and $d \neq 1$ then $T_{d+1,d}$ is empty.

If $\left(\frac{k}{5}\right) = 1$ and $d \neq 1$ then taking into account the isomorphism

$$o_5 \approx Z_5 \times Z_5,$$

the set $T_{d+1,d}$ consists of the pairs (u, w) such that mod 5,

$$w \equiv 1\ \text{and}\ u \equiv d \quad \text{or} \quad w \equiv d\ \text{and}\ u \equiv 1.$$

From this one finds from Theorem 7.1:

(3) $$\sum_{d=2}^{4} f(d) = \begin{cases} 6r^4 & \text{if } \left(\frac{k}{5}\right) = 1 \\ 0 & \text{if } \left(\frac{k}{5}\right) = 0 \\ 0 & \text{if } \left(\frac{k}{5}\right) = -1 \end{cases}$$

There remains to evaluate f(1). We have:

$$f(1) = \int_{0_5} h_{I+5M}(\text{Tr } z, Nz)\, dz \, ,$$

which was evaluated in formula (5) of §8. A final addition yields the values stated in the theorem.

Remark. In Theorem 11.3, with $r = 1/5$, we see that the first value, in case

$$\left(\frac{k}{5}\right) = 1$$

is substantially bigger (about twice as big) as the value in the two other cases. This means that when $k = Q(\sqrt{D})$ and

$$D \equiv 1, 4 \pmod{5}$$

one expects the frequency of occurrences of k to be correspondingly bigger. Indeed, in the tables, both the actual and predicted values for the occurrences of such k are big compared to the other cases, when $D \equiv 0, 2, 3 \pmod{5}$.

In Part III we shall determine the constant for the Serre group even when $K \cap k_{ab}$ is larger than Q_{ab} for a number of cases. We can apply this to $X_0(11)$ because the same type of argument as before will give the same factor $\frac{1}{5} \text{Num}_5(\tilde{\mathcal{G}})_5$. It is convenient to put this result here since it follows the same pattern that we just encountered.

Theorem 11.4. *Let* $k = Q(\sqrt{D})$ *and assume that* $5 \nmid D$. *Then*

$$\text{Num}_{330}(\tilde{\mathcal{G}}_{330}) = \frac{1}{5} \text{Num}_5(G_5, \mathfrak{o}_5^*) \, \text{Num}_{66}(\tilde{\mathcal{G}}_{66}) \, .$$

Proof. We assume that the reader is acquainted with the fibering terminology of Part III and the commutator manipulations of Part III, §3. Let L be a set of primes not containing 5 or 11. Then

$$G_{55L} = G_5 \times_{\phi_5} G_{11L}$$

$$G_{k,55L} = G_5 \times_{\phi_5} G_{k,11L}$$

whence

$$G'_{55L} = G'_5 \times G'_{11L}$$

$$G'_{k,55L} = G'_5 \times G'_{k,11L}.$$

In each of the cases considered, we have a field F such that F is abelian over k, with Galois group B, and

$$K \cap k_{ab} = Q_{ab}F.$$

Then $G_{330} = G_{30q}$ (with $q = 11$) is defined by a fibering

$$\phi = (\phi', \phi''),$$

$$\phi' : G_{30q} \longrightarrow \text{Gal}(F/k) \quad \text{and} \quad \phi'' : \prod_{\ell \mid 30q} \mathfrak{o}_\ell^* \longrightarrow \text{Gal}(F/k),$$

so that

$$\tilde{\mathcal{G}}_{30q} = G_{k,30q} \times_{N,\phi} \prod_{\ell \mid 30q} \mathfrak{o}_\ell^*$$

$$= (G_5 \times_{\phi_5} G_{k,6q}) \times_{N,\varphi} (\mathfrak{o}_5^* \times \mathfrak{o}_{6q}^*),$$

and

$$G_{30q} = \bigcup_\zeta G_{5,\zeta(5)} \times G_{11,\zeta} \times G_{2,\zeta(2)} \times G_3.$$

This gives

$$G_{k,30q} = \bigcup_{\zeta,\beta} [G_{5,\zeta(5)} \times (G_{11,\zeta} \times G_{2,\zeta(2)} \times G_3)_\beta],$$

where β ranges over B, and the index β follows our usual notation, denoting inverse image under ϕ. Therefore, we get

$$\bigcup_{\zeta,\beta} (G_{5,\zeta(5)} \times_N \mathfrak{o}_5^*) \times [(G_{11,\zeta} \times G_{2,\zeta(2)} \times G_3)_\beta \times_N (\mathfrak{o}_{11}^* \times \mathfrak{o}_2^* \times \mathfrak{o}_3^*)_\beta].$$

The use of the lemma concludes the proof.

PART III

SPECIAL COMPUTATIONS

In the second part, we had worked out a general formula for the constant giving the conjectured asymptotic behavior of Frobenius automorphisms in a GL_2-extension of the rationals K, with a representation of its Galois group

$$\rho' : G \longrightarrow \prod GL_2(Z_\ell) .$$

For all but a finite number of imaginary quadratic fields k, we had seen that

$$K \cap k_{ab} = Q_{ab} ,$$

and the constant was worked out explicitly in these cases.

We now come to the study of the exceptional quadratic fields such that $K \cap k_{ab} \neq Q_{ab}$. This requires special techniques of local class field theory, and is designed for individual curves which will be Serre curves, and the Shimura curve $X_0(11)$. The quadratic fields $Q(\sqrt{-1})$, $Q(\sqrt{-3})$, $Q(\sqrt{\Delta})$, $Q(\sqrt{-\Delta})$ play a special role.

The complications arise in these special cases because the exceptionally large intersection $K \cap k_{ab}$ forces us to evaluate the integrals giving the numerator of the constant by decomposing the domain of integration over fairly small sets, determined by the dependence relations of Galois and class field theory on this intersection.

For discriminants whose absolute value is < 100 we work out all cases except one, for our five curves, determining $K \cap k_{ab}$ and the corresponding constant. The single case we have not worked out, for the curve with $\Delta = -2^6 3^5$, $k = Q(\sqrt{-3})$, would have required additional complications. On the other hand, it seemed that it would be somewhat repetitive, without much additional insight arising from it, and was not worth the effort.

For those which we include, we already have to make an analysis of the manner in which the non-abelian operation of the matrices in the GL_2-extension corresponds to the abelian operation of the k-ideles. The most interesting case is that of §12. Applied to $X_0(11)$, it gives the theoretical explanation (conjecturally, of course) for the unusually large occurrence of the quadratic field

$k = \mathbb{Q}(\sqrt{-11})$, namely 88 times, which represents a confluence of several forces, including the fact that $-11 \equiv -1 \pmod 5$, cf. Theorem 11.3 of Part II, as well as the fact that $K \cap k_{ab} \neq \mathbb{Q}_{ab}$. Actually, the predicted value is a little high, see the comments on numerical results in Part IV.

An analogous situation arises for the curve with discriminant -43, and the field $k = \mathbb{Q}(\sqrt{-43})$ has a high frequency. In this case, the predicted value is perfectly in line with the actual count.

In all three cases $k = \mathbb{Q}(\sqrt{-3})$, $k = \mathbb{Q}(i)$, $k = \mathbb{Q}(\sqrt{\Delta})$, certain Galois symmetries give rise to simplifications in the computation of the desired integral for the numerator of the constant. The results are given in Theorems 7.3, 7.4, 7.5 for $\mathbb{Q}(\sqrt{-3})$, Theorem 9.1 for $\mathbb{Q}(i)$, and Theorem 12.1 for $\mathbb{Q}(\sqrt{\Delta})$. It turns out that for $\mathbb{Q}(i)$, the answer is the same as for the Serre fibering of Part II, §10. For $\mathbb{Q}(\sqrt{-3})$, it is the same in half the cases, and close to it in the other half. However, for $\mathbb{Q}(\sqrt{\Delta})$, there are substantial differences.

In the first sections, which we call general lemmas, we give a detailed series of lemmas on $GL_2(\mathbb{Z}_2)$ and $GL_2(\mathbb{Z}_3)$, especially concerning their subgroups of index 2, and the commutator subgroups, whose fixed fields are precisely the intersection $K \cap k_{ab}$. In §3 we determine those cases when $K \cap k_{ab} = \mathbb{Q}_{ab}$ which were not covered in Part II, because they required a somewhat finer argument than that contained in Part II, Theorem 3.1. We then go through systematically the fields $k = \mathbb{Q}(\sqrt{-3})$, $k = \mathbb{Q}(\sqrt{-1})$ and $k = \mathbb{Q}(\sqrt{\Delta})$.

PART III

SPECIAL COMPUTATIONS

GENERAL LEMMAS

1.	Lemmas on commutator subgroups	163
2.	$G_2 = GL_2(\mathbb{Z}_2)$	165
3.	Cases when $K \cap k_{ab} = \mathbb{Q}_{ab}$	174
4.	$K \cap k_{ab}$ when $k = \mathbb{Q}(\sqrt{-3})$ and $GL_2(\mathbb{Z}_3)$ splits	181
5.	$K \cap k_{ab}$ in other cases	185

$k = \mathbb{Q}(\sqrt{-3})$

6.	The action of $\mathcal{C}$ on $k(\Delta^{1/3})$	191
7.	The constant for Serre fiberings, $k = \mathbb{Q}(\sqrt{-3})$, $M = 2q$, q odd prime $\neq 3$, $\Delta = \pm q^n$	195
8.	Computation of integrals	201

$k = \mathbb{Q}(i)$

9.	The constant for Serre fiberings, q odd $\neq 3$	209

$k = \mathbb{Q}(\sqrt{\Delta})$

10.	The action of $\mathcal{C}$ on $k(A_2, \Delta^{1/4})$ when $k = \mathbb{Q}(\sqrt{\Delta})$	215
11.	The action of matrices on $k(A_4)$	218
12.	Computation of integrals and the constant	221

PART III

SPECIAL COMPUTATIONS

GENERAL FORMULAS

GENERAL LEMMAS

§1. Lemmas on commutator subgroups

Lemma 1. *Let* q *be a prime number. Let* $r \geq 1$. *Let*

$$W_{q,r} = W_r = I + q^r M_q.$$

Then

$$W'_r = W_{2r} \cap SL_2(Z_q).$$

Proof. We write a commutator from $I + q^r M_q$,

$$(I+q^r X)(I+q^r Y)(I-q^r X+q^{2r} X^2)(I-q^r Y+q^{2r} Y^2) \quad (\mod q^{2r+1})$$

$$= I + q^{2r}(XY - YX) \quad (\mod q^{2r+1}).$$

It is easy to see that the vector space over F_q generated by the matrices of the form

$$XY - YX$$

has dimension 3, and consists of the matrices with trace 0. Hence W'_r is the unique closed subgroup of W_{2r} having this space as associated vector space at level $2r$. Furthermore, W'_r is contained in $SL_2(Z_q)$, and the associated vector space of

$$W_{2r} \cap SL_2(Z_q)$$

at level $2r$ has dimension ≤ 3. This proves the desired equality.

Lemma 2. *Let* q *be a prime* ≥ 3. *Let* r *denote reduction* mod q. *Let*

$$V = r^{-1}(SL_q(q)) = (I+qM_q)SL_2(Z_q).$$

Then

$$V/V' \approx SL_2(F_q)/SL_2(F_q)',$$

and

$$V' = r^{-1}(SL_2(F_q)') \cap SL_2(Z_q).$$

Proof. There are commutators from V of the form

$$\sigma(I+qX)\sigma^{-1}(I-qX) = I + q(X - \sigma X\sigma^{-1}) \quad (\mathrm{mod}\ q^2),$$

and $\sigma X \sigma^{-1}$ depends only on σ mod q. It is easy to get three linearly independent matrices mod q, out of the expression

$$X - \sigma X \sigma^{-1},$$

where X has trace 0 mod q. This shows that the associated vector space in M_q/qM_q to the closed subgroup V' consists of the matrices of trace 0. On the other hand, the closed subgroup of $I + qM_q$ belonging to this space is

$$(I+qM_q) \cap SL_2(Z_q).$$

The lemma follows at once.

§2. $G_2 = GL_2(\mathbb{Z}_2)$

In this section we gather together mostly group theoretic facts about $GL_2(\mathbb{Z}_2)$, which we denote by G_2.

We are interested first in the (closed) subgroups of index 2. They correspond to characters of order 2. If need be, any group of order 2 is identified with $\{\pm 1\}$.

Suppose that the character factors through the determinant homomorphism, and hence amounts to a homomorphism of

$$\mathbb{Z}_2^*/\mathbb{Z}_2^{*2} \approx Z(8)^*.$$

Note that 1, 3, 5, 7 mod 8 represent the elements of $Z(8)^*$. We have three possible characters, denoted by χ_2, χ_{-2}, χ_i such that

$$\text{Ker } \chi_2 = \{1,7\}, \qquad \text{Ker } \chi_{-2} = \{1,3\}, \qquad \text{Ker } \chi_i = \{1,5\}.$$

If G_2 is the Galois group of an extension of $\mathbb{Q}$, inducing the determinant character on the roots of unity, then the indices 2, -2, i indicate the quadratic field fixed by the kernel of the character, namely

$$\mathbb{Q}(\sqrt{2}), \qquad \mathbb{Q}(\sqrt{-2}), \qquad \mathbb{Q}(i)$$

respectively.

On the other hand we have the sign homomorphism

$$\varepsilon : GL_2(\mathbb{Z}_2) \longrightarrow GL(2) \approx S_3 \longrightarrow \{\pm 1\},$$

where S_3 is the permutation group. We then obtain four subgroups

$$E_2 = \text{Ker } \varepsilon, \qquad \text{Ker } \varepsilon\chi_2, \qquad \text{Ker } \varepsilon\chi_{-2}, \qquad \text{Ker } \varepsilon\chi_i.$$

Lemma 1. *The above subgroups of index 2 are the only ones. The subgroups*

$$\text{Ker } \chi_2, \qquad \text{Ker } \chi_{-2}, \qquad \text{Ker } \varepsilon\chi_2, \qquad \text{Ker } \varepsilon\chi_{-2}$$

do not contain $I + 4M_2$. *The other three subgroups contain* $I + 4M_2$.

Proof. The first statement will follow from the determination of $GL_2(Z_2)'$ in Theorem 2.1. The characterization of the kernels in terms of $I + 4M_2$ is immediate and left to the reader.

Theorem 2.1. *The homomorphism* $\varepsilon \times \det$ *is the maximal abelianizing homomorphism of* $GL_2(Z_2)$. *In other words*,

$$GL_2(Z_2)' = E_2 \cap SL_2(Z_2).$$

Proof. Since E_2 has index 2 in G_2 it suffices to prove that

$$(SL_2(Z_2) : G_2') = 2.$$

For the rest of this section, as we deal only at the prime 2, we abbreviate:

$$E_2 = E, \qquad G_2 = G, \qquad M_2 = M.$$

To do what we want, we now see that it suffices to do it mod 4.

Lemma 2.

(i) $\qquad\qquad\qquad I + 2M \subset E.$

(ii) $\qquad\qquad G' \supset (I+2M)' = (I+4M) \cap SL_2(Z_2)$

(iii) $\qquad\qquad\qquad (I+4M) \cap G' \subset E'$

(iv) *Reduction* mod 4 *gives an isomorphism*
$$SL_2(Z_2)/G' \approx SL_2(4)/G(4)'.$$

Proof. That $I + 2M \subset E$ is obvious. Lemma 2 of §1, applied to the prime 2, yields (ii). Hence the inverse image of $G(4)'$ in $SL_2(Z_2)$ is G'. The rest of the lemma then follows obviously.

Lemma 3. *Let* $2M(4)_0$ *be the set of matrices in* $2M(4)$ *with trace* 0. *Then* $I + 2M(4)_0$ *is contained in* $G(4)'$.

Proof. Let

$$\gamma = \begin{pmatrix} -1 & -1 \\ 1 & 0 \end{pmatrix} \quad \text{and} \quad \beta = \begin{pmatrix} 0 & 1 \\ 1 & 0 \end{pmatrix}.$$

Forming commutators of elements of type $I + 2A$ with γ and β yields elements of the form
$$I + 2(A - \gamma A \gamma^{-1}) \quad \text{and} \quad I + 2(A - \beta A \beta^{-1}) \mod 4.$$

It is immediately seen that the associated vector space of such elements in $2M/4M$ has dimension 3, in other words, is the space of matrices with trace 0, as desired.

Returning to the theorem, via Lemma 2, we consider the sequence of subgroups
$$G(4) \supset SL_2(4) \supset G(4)'.$$

It is clear that $SL_2(4)$ has index 2 in $G(4)$. Note that $G(4)'$ has $G(2)'$ as factor group, and that $G(2)'$ is the commutator group of $G(2) \approx S_3$, and has order 3. The group
$$I + 2M(4)_0$$
has order 8, and is contained in the kernel of the reduction mod 2:
$$G(4)' \longrightarrow G(2)',$$
so that $G(4)'$ has order at least $3 \cdot 8 = 24$. But $G(4)$ has order $4 \cdot 24$. Hence
$$(G(4) : G(4)') \leq 4.$$

Since $G(4)' \neq SL_2(4)$ (because of the existence of ε), it follows that the index is exactly 4, and also that
$$(SL_2(4) : G(4)') = 2,$$
thereby proving the theorem.

We can also formulate Lemma 3 in a 2-adic way.

Theorem 2.2. $GL_2(Z_2)' \cap (I + 2M_2) = SL_2(Z_2) \cap (I + 2M_2)$.

Proof. We have seen that this is true mod 4, and the result follows by refinement because of Lemma 2, (ii) (after level 4, the group has the right associated vector space mod higher powers of 2).

The next theorem determines V' when V is any one of the subgroups of $GL_2(Z_2)$ not containing $I + 4M$.

Theorem 2.3. *Let* V *be any one of the subgroups of* $GL_2(\mathbb{Z}_2)$ *of index* 2, *which does not contain* $I + 4M$, *i.e.*

$$\text{Ker } \chi_2, \qquad \text{Ker } \chi_{-2}, \qquad \text{Ker } \varepsilon\chi_2, \qquad \text{Ker } \varepsilon\chi_{-2}.$$

Then

$$V' = GL_2(\mathbb{Z}_2)'.$$

Proof. By hypothesis,

$$V(I + 4M_2) = GL_2(\mathbb{Z}_2).$$

Hence we obtain an isomorphism

$$V/[V \cap (I + 4M_2)] \approx GL_2(\mathbb{Z}_2)/(I + 4M_2).$$

On the other hand, if Y is a normal subgroup of X, then we have the formula

$$(X/Y)' = X'/(X' \cap Y).$$

Consequently we obtain an isomorphism

$$V'/[V' \cap (I + 4M_2)] \approx GL_2(\mathbb{Z}_2)'/[GL_2(\mathbb{Z}_2)' \cap (I + 4M_2)].$$

Since $GL_2(\mathbb{Z}_2)' \subset SL_2(\mathbb{Z}_2)$, it suffices to prove:

Lemma 5. *Let* V *be any one of the subgroups of* $GL_2(\mathbb{Z}_2)$ *which does not contain* $I + 4M$. *Then*

$$V' \cap (I + 4M_2) = SL_2(\mathbb{Z}_2) \cap (I + 4M_2).$$

Proof. Note that

$$V' \cap (I + 4M_2) \supset [V \cap (I + 2M_2)]'.$$

For matrices A, B the commutator of $I + 2A$ and $I + 2B$ is

$$I + 4(AB - BA) \bmod 8.$$

We can pick matrices A, B such that $I + 2A$ and $I + 2B$ lie in

$$V \cap (I + 2M_2)$$

and such that their commutators give enough elements to show that the vector subspace of $4M_2/8M_2$ associated with the closed subgroup

$$[V \cap (I+2M_2)]'$$

is the space of matrices with trace 0. Since this subspace determines the associated group uniquely, the lemma will be proved. There remains to exhibit the matrices A and B.

For χ_{-2} and $\varepsilon\chi_{-2}$ we use for instance:

$$A = \begin{pmatrix} 0 & 1 \\ 0 & 0 \end{pmatrix} \quad \text{and} \quad B = \begin{pmatrix} 0 & 0 \\ 1 & 0 \end{pmatrix}, \quad AB-BA = \begin{pmatrix} 1 & 0 \\ 0 & -1 \end{pmatrix}$$

$$A = \begin{pmatrix} 1 & 0 \\ 0 & 0 \end{pmatrix} \quad B = \begin{pmatrix} 0 & 1 \\ 0 & 0 \end{pmatrix}, \quad AB-BA = \begin{pmatrix} 0 & 1 \\ 1 & 0 \end{pmatrix}$$

$$A = \begin{pmatrix} 1 & 0 \\ 0 & 0 \end{pmatrix} \quad B = \begin{pmatrix} 0 & 0 \\ 1 & 0 \end{pmatrix}, \quad AB-BA = \begin{pmatrix} 0 & 0 \\ 1 & 0 \end{pmatrix}.$$

For χ_2 and $\varepsilon\chi_2$ we take the same A, B in the first line, but use $-A$ instead of A in the second and third line. This concludes the proof.

Next we handle $\operatorname{Ker}\varepsilon = E$ and $\operatorname{Ker}\chi_i$. We shall not treat $\operatorname{Ker}\varepsilon\chi_i$, which is a little more complicated, and is not needed in the range of computations which we worked out. Whereas we shall see that we can treat the first two groups mod 4, it would require going to congruences mod 8 to treat $\operatorname{Ker}\varepsilon\chi_i$ in an analogous way.

Let
$$SW_{2,2} = (I+4M) \cap SL_2(\mathbb{Z}_2).$$

Then if $V = E = \operatorname{Ker}\varepsilon$, or $\operatorname{Ker}\chi_i$, we shall prove that $V' \supset SW_{2,2}$, and we shall obtain a diagram where the horizontal arrows induce an isomorphism of factor groups at each level.

$$\begin{array}{ccc} SL_2(\mathbb{Z}_2) & \longrightarrow & SL_2(4) \\ | & & | \\ G' & \longrightarrow & G'(4) \\ | & & | \\ V' & \longrightarrow & V'(4) \\ | & & | \\ SW_{2,2} & \longrightarrow & 1 \end{array}$$

Each group on the left is the inverse image of the corresponding group on the right.

The even and odd matrices mod 2 play a pervasive role. We remind the reader of what they look like:

$$O(2): \begin{pmatrix} 0 & 1 \\ 1 & 0 \end{pmatrix}, \quad \begin{pmatrix} 1 & 1 \\ 0 & 1 \end{pmatrix}, \quad \begin{pmatrix} 1 & 0 \\ 1 & 1 \end{pmatrix}$$

$$E(2): \begin{pmatrix} 1 & 0 \\ 0 & 1 \end{pmatrix}, \quad \begin{pmatrix} 0 & 1 \\ 1 & 1 \end{pmatrix}, \quad \begin{pmatrix} 1 & 1 \\ 1 & 0 \end{pmatrix}.$$

We note that $O(2)$ and the zero matrix form an additive group, and also $E(2)$ and the zero matrix form an additive group. Therefore

$$I + 2O(2) \cup I \pmod 4 \quad \text{and} \quad I + 2E(2) \cup I \pmod 4$$

form multiplicative subgroups of $GL_2(4) = G(4)$.

Theorem 2.4.

(i) *The group E is generated by*

$$I + 2M \quad \text{and} \quad \gamma = \begin{pmatrix} -1 & -1 \\ 1 & 0 \end{pmatrix}.$$

(ii) $E' \supset SW_{2,2}$.

(iii) $E'(4) = (I + 2O(2)) \cup I$.

(iv) *Reduction mod 4 gives isomorphisms*

$$G'/E' \approx G(4)'/E(4)' \quad \text{and} \quad E'/SW_{2,2} \approx E'(4).$$

Proof. Since E contains $I + 2M$, and γ has order 3 (mod 2), the first assertion is obvious. The second repeats part of Lemma 2, (ii). The third is obtained by computing commutators in a trivial way. Then (iv) follows from Lemma 2 (iii), as was to be shown.

Theorem 2.5. *Let* $V = \text{Ker } \chi_i$. *Then:*

(i) $V = SL_2(Z_2)(I + 4M)$.

(ii) $V' \supset SW_{2,2}$.

(iii) $V'(4) = SL_2(4)'$ *has order* 12. *It is generated by*

$$I + 2O(2)$$

and by any element of order 3 in G(4). It contains all the elements of order 3 in G(4).

(iv) Reduction mod 4 gives an isomorphism

$$V'/SW_{2,2} \approx V'(4).$$

Proof. From $I+4M \subset V$ and Lemma 1 of §1 we conclude that

$$(I+8M) \cap SL_2(Z_2) \subset V'.$$

Statement (ii) now amounts to showing that the associated vector space of V' in $4M/8M$ has dimension 3, i.e. consists of the elements with trace 0. For this we have to find enough commutators.

Pick first

$$A = \begin{pmatrix} 0 & 1 \\ 0 & 0 \end{pmatrix} \quad \text{and} \quad B = \begin{pmatrix} 0 & 0 \\ 1 & 0 \end{pmatrix}.$$

Then

(1) $$[I+2A, I+2B] \equiv I + 4\begin{pmatrix} 1 & 0 \\ 0 & 1 \end{pmatrix} \pmod{8}.$$

We then form commutators

$$[I+2A, \sigma] \equiv I + 2(A - \sigma A \sigma^{-1}) \pmod{4}$$

with two cases:

(2) $$A = \begin{pmatrix} 1 & 1 \\ 0 & 1 \end{pmatrix} \quad \text{and} \quad \sigma = \begin{pmatrix} 0 & 1 \\ -1 & 0 \end{pmatrix}, \quad [I+2A, \sigma] \equiv I + 2\begin{pmatrix} 0 & 1 \\ -1 & 0 \end{pmatrix}$$

(3) $$A = \begin{pmatrix} 0 & 1 \\ 0 & 0 \end{pmatrix} \quad \text{and} \quad \sigma = \begin{pmatrix} -1 & 1 \\ -1 & 0 \end{pmatrix}, \quad [I+2A, \sigma] \equiv I + 2\begin{pmatrix} 1 & 1 \\ 1 & -1 \end{pmatrix}$$

Squaring the commutators in the second and third case yields

$$I + 4\begin{pmatrix} -1 & 1 \\ -1 & -1 \end{pmatrix} \quad \text{and} \quad I + 4\begin{pmatrix} 2 & 0 \\ 1 & 0 \end{pmatrix}, \pmod{8}.$$

This gives us the 3-dimensional space of trace 0 in $4M/8M$.

We also see that $V'(4)$ contains $I + 2O(2)$. Pick any element of order 3, say γ, and form its commutator with

$$\begin{pmatrix} 0 & 1 \\ -1 & 0 \end{pmatrix}.$$

We find an element of order 3, which lies in $V'(4)$. Hence $V'(4)$ has order at least 12. Let H be the subgroup generated by

$$I + 2O(2)$$

and by an element of order 3 in $V'(4)$. Then one sees that $-I \notin H$, and H is of index 2 in $V(4)/\pm I$, whence normal, and also normal in $V(4)$ since $-I$ lies in the center. The factor group is abelian, and hence $H \supset V'(4)$. Therefore $H = V'(4)$. It now follows that $V'(4)$ has order 12, and $V'(4) = SL_2(4)'$, so that $V'(4)$ is normal in $G(4)$. The rest of (iii) follows at once, because the 3-Sylow groups are conjugate, and (iv) is clear, thereby proving the theorem.

The purely group theoretic arguments which precede are used in the context when $G(4) = GL_2(4)$ is the Galois group of the 4-division points of an elliptic curve A, i.e.

$$G(4) = \text{Gal}(Q(A_4)/Q).$$

It is a classical fact that $\Delta^{1/4}$ is contained in the field of 4-division points. The corresponding lattice of fields for Theorems 2.4 and 2.5 can then be drawn as on the figure. The small numbers near each line indicate the degree of the corresponding field extension.

The field $Q(x(A_4))$ is the field of x-coordinates of the 4-division points – we could call it the **modular subfield** of 4-division points.

Since for the moment we only need the existence of $\Delta^{1/4}$ in the field of 4-division points, we don't go any further into the matter. In §11, we shall prove this fact, along with more precise information on how the matrices and idele class groups operate on $Q(A_4)$.

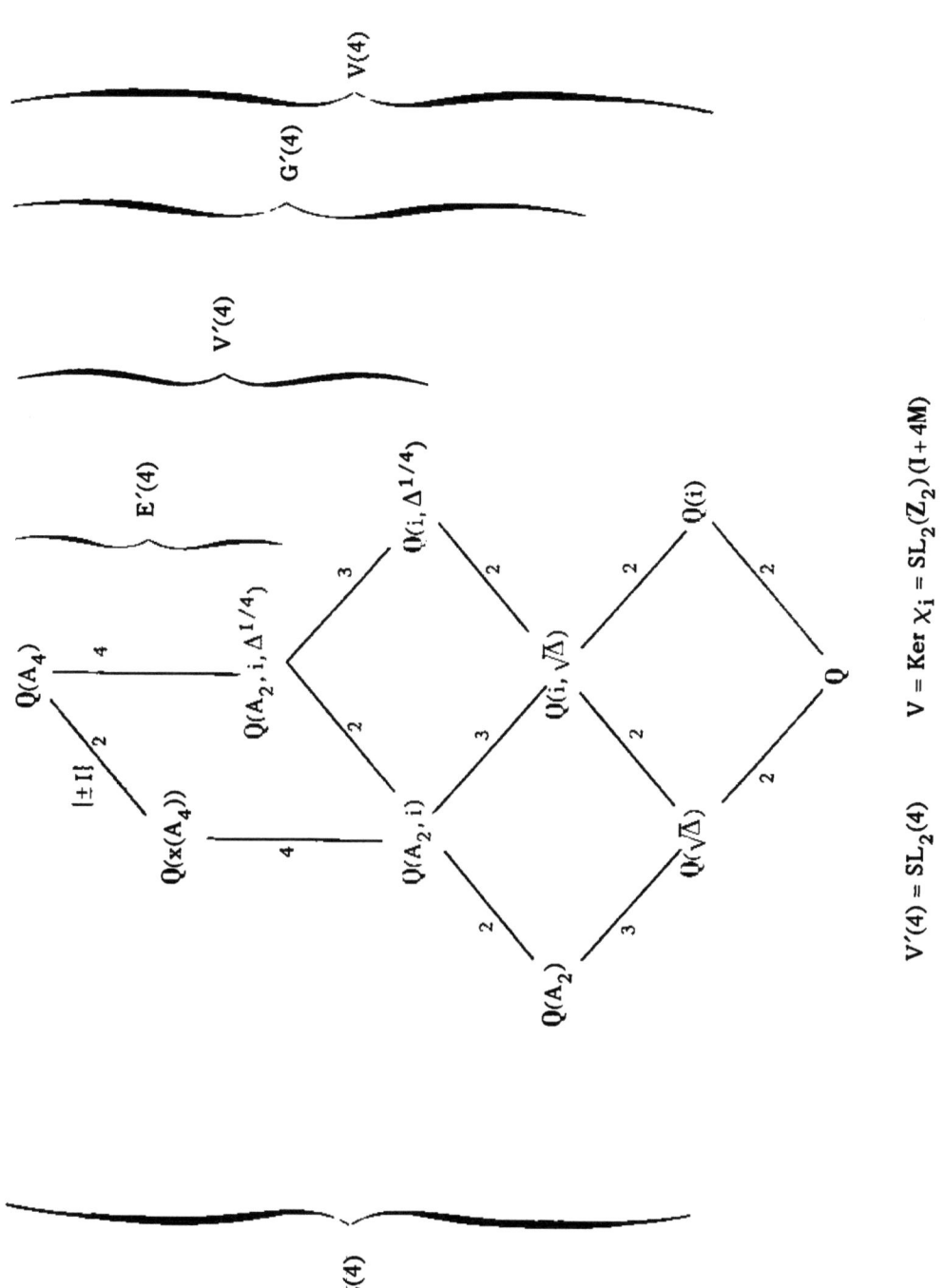

§3. Cases when $K \cap k_{ab} = Q_{ab}$

In Part II, §3, we gave a table of those cases when $K \cap k_{ab}$ is equal to Q_{ab}. We now want to prove the results justifying the table. We begin with lemmas of general group theory.

Let G_1, G_2 be groups, and let B be an abelian group. Let

$$\phi_1 : G_1 \longrightarrow B \quad \text{and} \quad \phi_2 : G_2 \longrightarrow B$$

be surjective homomorphisms. We define the fiber product over ϕ,

$$G = G_1 \times_{(\phi_1, \phi_2)} G_2 = G_1 \times_\phi G_2$$

to be the set of pairs (σ_1, σ_2) such that $\phi_1(\sigma_1) = \phi_2(\sigma_2)$. We say that the fibering ϕ **dissolves under commutators** if

$$G' = G'_1 \times G'_2 \, .$$

Lemma 1. *Let* $G = G_1 \times_\phi G_2$ *be a fiber product as above. Assume that* $\phi_2^{-1}(B)$ *contains a subset of elements which map onto* B *under* ϕ_2 *and which commute with each other. Then*

$$G' = G'_1 \times G'_2 \, ,$$

i.e. the fibering dissolves under commutators.

Proof. We have $\mathrm{pr}_1 \, G' = G'_1$ and $\mathrm{pr}_2 \, G' = G'_2$ because the first and second projections are surjective. It suffices therefore to prove that G' contains $G'_1 \times \{e_2\}$. Given $\sigma_1, \tau_1 \in G_1$, we can find $\sigma_2, \tau_2 \in G_2$ which commute with each other, and such that

$$\phi_1(\sigma_1) = \phi_2(\sigma_2), \qquad \phi_1(\tau_1) = \phi_2(\tau_2) \, .$$

The commutator of (σ_1, σ_2) and (τ_1, τ_2) gives what we want.

Remarks. The proof shows that the lemma holds under the weaker hypothesis that given a pair of elements $\sigma_1, \tau_1 \in G_1$ we can find σ_2, τ_2 in G_2 commuting with each other, and having the same images under ϕ_1, ϕ_2 respectively.

The condition under which the lemma holds is obviously satisfied if the group B is cyclic.

We apply the lemma when G is the Galois group of our usual GL_2-extension. We need more systematic notation for a set of primes and its complement. If M is a positive integer, we let [M] denote the set of primes not dividing M. Then

$$G \subset G_M \times G_{[M]}.$$

Let M, L be disjoint sets of primes. Suppose that $Q(\sqrt{\Delta})$ is contained in the fields K_M and K_L, whose Galois groups are G_M and G_L respectively. We write

$$G_M \times_\Delta G_L$$

for the fibered product with respect to the homomorphisms

$$\phi_1 : G_M \longrightarrow \text{Gal}(Q(\sqrt{\Delta})/Q) \quad \text{and} \quad \phi_2 : G_L \longrightarrow \text{Gal}(Q(\sqrt{\Delta})/Q).$$

We call this **Serre's fibering.**

As we have already seen in dealing with Serre subgroups, we always have

$$\sqrt{\Delta} \in K_2.$$

We denote by Δ_0 the discriminant of the field $Q(\sqrt{\Delta})$. We shall be specifically interested in the case when

$$\Delta_0 = \pm q$$

where q is an odd prime, since all the five curves which we consider have this property. In this case, only q can ramify in $Q(\sqrt{\Delta})$, and consequently we also have

$$Q(\sqrt{\Delta}) \subset K_q.$$

Let k be a quadratic imaginary extension of Q as usual. As before, $G_k = \text{Gal}(K/k)$. We first give a result concerning G'/G'_k.

Theorem 3.1. *Let q be an odd prime and let $k = Q(\sqrt{\Delta})$. Assume:*

(i) $\Delta_0 = -q$

(ii) $G = GL_2(\mathbb{Z}_2) \times_\Delta G_{[2]}$.

(iii) $G_{[2]} = GL_2(\mathbb{Z}_q) \times_\phi G_{[2q]}$ is a cyclic fibering.

Then
$$G'/G'_k = GL_2(\mathbb{Z}_2)'/E'_2 \times G'_q/E'_q.$$

If $q \geq 5$, then the term G'_q/E'_q is 1, and can be omitted.

Proof. First we have by Lemma 1,

(1) $$G' = G'_2 \times G'_q \times G'_{[2q]}.$$

Next, we have

(2) $$G_k = E_2 \times G_{[2],k},$$

because the left hand side and right hand side of this expression both have index 4 in $G_2 \times G_{[2]}$, and the right hand side is obviously contained in the left hand side. Since q is the only ramified prime in k, we also have

(3) $$G_{[2],k} = E_q \times_\phi G_{[2q]},$$

for some cyclic fibration ϕ, which is in fact of order 2. Taking commutators and using Lemma 1, we find

(4) $$G'_k = E'_2 \times E'_q \times G'_{[2q]}.$$

The theorem follows at once, because $SL_2(\mathbb{Z}_q) \subset E_q$, and for $q \geq 5$, we know from Part II, Lemma 1 of §3 that $SL_2(\mathbb{Z}_q)' = SL_2(\mathbb{Z}_q)$.

Remark. The hypotheses of Theorem 3.1 apply to a Serre curve, with Galois group
$$G = S_{2q} \times G_{[2q]}.$$

According to Part I, §8, they also apply to $X_0(11)$.

Theorem 3.2. Let $k = \mathbb{Q}(\sqrt{-8})$. Assume that
$$G = GL_2(\mathbb{Z}_2) \times_\phi G_{[2]}$$
is a fibering where $\phi = (\varepsilon, \phi_{[2]})$. Then
$$G' = G'_k, \quad \text{that is} \quad K \cap k_{ab} = \mathbb{Q}_{ab}.$$

Proof. We first note that $k \subset K_2$, and so

$$G_k = G_{k,2} \times_\phi G_{[2]}.$$

Hence by Lemma 1,

$$G'_k = G'_{k,2} \times G'_{[2]}$$

and also

$$G' = GL_2(\mathbb{Z}_2)' \times G'_{[2]}.$$

The theorem follows from Theorem 2.3, with $V = G_{k,2}$.

Remark. In the preceding theorem, and also the two subsequent ones, the essential factors are G_2 and G_q, with a possible further fibering between G_q and $G_{[2q]}$. As we have already encountered in studying $X_0(11)$, it is useful to consider only partial products G_L instead of $G_{[2q]}$, where L is a set of primes not containing 2 or q. All these theorems apply mutas mutandis to this case. Stating them with the added L into the notation gets heavy, and it seemed preferable to make the remark instead.

Any discriminant of a quadratic field can be factored uniquely into a product of discriminants each of which has only one prime factor. If D_1, D_2 are discriminants, and also $D_1 D_2$, then we let

$$\chi_{D_1 D_2} = \chi_{D_1} \chi_{D_2}$$

be the character associated with the corresponding quadratic field. They factor through the determinant homomorphism, and are defined on $\mathbb{Z}^*_{D_1 D_2}$.

Theorem 3.3. Let $k = Q(\sqrt{-24})$. Assume that $\Delta_0 = \pm q$ where q is an odd prime. Assume furthermore:

(i) $G = GL_2(\mathbb{Z}_2) \times_\Delta G_{[2]}$

(ii) G_3 splits in $G_{[2]}$.

Then $G' = G'_k$.

Proof. We distinguish two cases.

Case 1. $q \neq 3$.

We can write -24 as a product of discriminants,

$$-24 = 8(-3).$$

Then

$$G_k = GL_2(\mathbb{Z}_2) \times_\phi G_{[2]},$$

where $\phi = (\phi_1, \phi_2)$ is a homomorphism into $Z(2) \times Z(2)$, and

$$\phi_1 = \varepsilon \times \chi_2, \qquad \phi_2 = \chi_{q^*} \times \chi_{-3}.$$

As usual, $q^* = \pm q$, taking the sign which makes $q^* \equiv 1 \pmod 4$. Since $q \neq 3$, the characters χ_{q^*} and χ_{-3} factor through distinct factors of $G_{[2]}$, because we assumed that G_3 splits. Hence ϕ_2 satisfies the condition of Lemma 1, and ϕ dissolves under commutators. So does the fibering for G itself, and the theorem follows.

Case 2. $q = 3$.

Then $\Delta_0 = -3$ since Δ_0 is a discriminant. We can write

$$G = (GL_2(\mathbb{Z}_2) \times_\Delta G_3) \times G_{[6]}$$

and

$$G_k = (G_{k,2} \times_\Delta G_3) \times G_{[6]},$$

where

$$G_{k,2} = \operatorname{Ker} \varepsilon \chi_2.$$

Then Lemma 1 gives

$$G' = GL_2(\mathbb{Z}_2)' \times G'_3 \times G'_{[6]}$$

$$G'_k = G'_{k,2} \times G'_3 \times G'_{[6]}.$$

Hence

$$G'/G'_k = GL_2(\mathbb{Z}_2)'/G'_{k,2}.$$

We let $V = G_{k,2}$, and apply Theorem 2.3 to conclude the proof.

Theorem 3.4. *Let* $k = \mathbb{Q}(\sqrt{-8q})$ *where* q *is an odd prime, and* $D = -8q$. *Assume that* $\Delta_0 = \pm q$. *Assume further that*

$$G = GL_2(\mathbb{Z}_2) \times_\Delta G_q \times_\phi G_{[2q]}$$

where ϕ is a fibering which dissolves under commutators, between G_q and $G_{[2q]}$. Then
$$G' = G'_k.$$

Proof. We have
$$G_k = (G_{k,2} \times_\Delta G_q) \times_\phi G_{[2q]}$$
and
$$G_{k,2} = \text{Ker } \varepsilon \times_{D/\Delta_0}.$$
Then
$$G' = GL_2(\mathbb{Z}_2)' \times G'_q \times G'_{[2q]}$$
and
$$G'_k = G'_{k,2} \times G'_q \times G'_{[2q]}$$

by the associativity of the fiberings, because the fibering over Δ is only between $GL_2(\mathbb{Z}_2)$ and G_q under the present hypotheses. We put $V = G_{k,2}$ and use Theorem 2.3 to conclude the proof.

Theorem 3.5. *Let* $k = Q(\sqrt{D})$ *and suppose that* $5|D$. *Assume that*
$$G = G_5 \times_\phi G_{[5]},$$
and that ϕ *is cyclic of order* 5. *Then*
$$G' = G'_k.$$

Proof. Since D is a discriminant, we cannot have $D = -5$, and hence some other prime divides D. Therefore
$$G_k = G_5 \times_{\phi,D} G_{[5]},$$
and the fibering for G_k is cyclic of order 10, so dissolves under commutators. The theorem follows at once.

The assumptions of Theorems 3.4 and 3.5 are of course designed for application to $X_0(11)$.

Again consider the case which will be applied to $X_0(11)$, namely assume that

$$G = G_5 \times_\phi G_{[5]},$$

and assume that ϕ dissolves under commutators.

Let $k = \mathbb{Q}(\sqrt{D})$, and suppose that $5 \nmid D$. Then

$$G_k = G_5 \times_\phi G_{[5],k},$$

and consequently

$$G'_k = G'_{[5]}/G'_{[5],k}.$$

This gets rid of the factor at 5, and as we already mentioned in a remark, we conclude:

Theorem 3.6. *Let* G *be the Galois group of division points of* $X_0(11)$. *If* k *is any one of the fields with discriminant*

$$D = -8, -15, -20, -24, -40, -55, -88,$$

then $G' = G'_k$, *that is,*

$$K \cap k_{ab} = \mathbb{Q}_{ab}.$$

Proof. We know from Part I, Theorem 8.3, that

$$G_{[5]} = S_{22} \times \prod_{\ell \neq 2,5,11} GL_2(\mathbb{Z}_\ell)$$

so the Serre curve results apply.

§4. $K \cap k_{ab}$ when $k = Q(\sqrt{-3})$ and $GL_2(Z_3)$ splits

Throughout this section, we let $k = Q(\sqrt{-3})$.
We suppose that the representation

$$\rho': G \longrightarrow \prod GL_2(Z_\ell)$$

is that associated with an elliptic curve A with discriminant Δ. It is a classical fact that $\Delta^{1/3}$ is contained in the field $Q(A_3)$ of division points of order 3. Indeed, one can see this from the fact that $j^{1/3}$ is a modular function of level 3 (see for instance [L 1], Chapter 18, §5, Theorem 8), and

$$j^{1/3} = g_2/\Delta^{1/3} .$$

The field generated by the x-coordinates (Weber-coordinate) of the points of order 3 is the same as the field generated by the values of the modular functions of level 3, cf. [L 1], Chapter 9, §3. Since $k = Q(\sqrt{-3})$ we see that

$$k(\Delta^{1/3})$$

is abelian, and in fact cyclic over k.

If the Galois group of $Q(A_3)$ over Q is all of $GL_2(3)$, then it follows that Δ is not a rational cube, and the above cyclic extension has precise degree 3. The subgroup of $GL_2(3)$ leaving k fixed is precisely $SL_2(3)$.

For future reference, we recall some facts concerning $GL_2(3)$. It operates as a permutation group of the subspaces of dimension 1 of F_3^2, and with this operation we have an isomorphism

$$GL_2(3)/\pm 1 \approx S_4 .$$

This induces an isomorphism

$$SL_2(3)/\pm 1 \approx A_4$$

with the alternating group, which has 12 elements. It is then easy to verify that the factor commutator group is given by

$$A_4/A_4' = Z(3) .$$

In fact, one can display explicitly the commutator group $SL_2(3)'$ and its cosets in $SL_2(3)$ as follows:

$$C_0 = SL_2(3)' = \left\{ \pm I, \pm \begin{pmatrix} 0 & 1 \\ -1 & 0 \end{pmatrix}, \pm \begin{pmatrix} 1 & 1 \\ 1 & -1 \end{pmatrix}, \pm \begin{pmatrix} -1 & 1 \\ 1 & 1 \end{pmatrix} \right\}$$

$$C_1 = \left\{ \pm \begin{pmatrix} 1 & 1 \\ 0 & 1 \end{pmatrix}, \pm \begin{pmatrix} 0 & 1 \\ -1 & -1 \end{pmatrix}, \pm \begin{pmatrix} 1 & -1 \\ 1 & 0 \end{pmatrix}, \pm \begin{pmatrix} 1 & 0 \\ -1 & 1 \end{pmatrix} \right\}$$

$$C_2 = \left\{ \pm \begin{pmatrix} 1 & -1 \\ 0 & 1 \end{pmatrix}, \pm \begin{pmatrix} 0 & 1 \\ -1 & 1 \end{pmatrix}, \pm \begin{pmatrix} 1 & 0 \\ 1 & 1 \end{pmatrix}, \pm \begin{pmatrix} -1 & -1 \\ 1 & 0 \end{pmatrix} \right\}.$$

As a matter of notation, we let

$$E_3 = SL_2(\mathbb{Z}_3)(I+3M_3).$$

Then

$$SL_2(3) = E_3(3).$$

The elements of the three cosets above can be characterized as follows, in $SL_2(3)$.

$$SL_2'(3) = E_3'(3) = \begin{cases} I \text{ and } -I \\ \text{all elements with trace } 0 \text{ mod } 3 \end{cases}$$

The two cosets in $SL_2(3) = \begin{cases} \text{elements} \neq I \text{ or } -I \\ \text{all non scalar mod } 3 \\ \text{all elements with trace} \neq 0 \text{ mod } 3. \end{cases}$

Lemma 1. *Let* r *be reduction* mod 3. *Then*

$$E_3' = r^{-1}(SL_2(3)') \cap SL_2(\mathbb{Z}_3)$$

and we have an isomorphism

$$SL_2(\mathbb{Z}_3)/E_3' \approx SL_2(3)/SL_2(3)'.$$

In particular,

$$(SL_2(\mathbb{Z}_3) : E_3') = 3.$$

Proof. The first part of the lemma is independent of the prime 3 and was proved as Lemma 2 of §1. The second part follows from the preceding remarks on $GL_2(\mathbb{Z}_3)$.

The corresponding field diagram on the field of 3-division points can be drawn as follows.

$$E_3(3) = SL_2(3) \left\{ \begin{array}{c} k(A_3) \\ | \\ k(\Delta^{1/3}) \\ | \\ k \end{array} \right\} E_3'(3)$$

Note that $GL_2'(3) = SL_2(3)$ and $GL_2'(\mathbb{Z}_3) = SL_2(\mathbb{Z}_3)$.

The preceding purely group theoretic lemma can be applied in the concrete situation of division points. In our notation, we let $G_k = \text{Gal}(K/k)$, and

$$G_{k,3} = \text{Gal}(K_3/k) = \text{Gal}(Q(A^{(3)})/k).$$

Theorem 4.1.
(i) If $G_3 = GL_2(\mathbb{Z}_3)$ then $G_{k,3} = E_3$.
(ii) If $G_{k,3} = E_3$ then
$$K_3 \cap k_{ab} = Q_{ab,3}(\Delta^{1/3}).$$

Proof. The first assertion is merely a lifting 3-adically of the statement already made that $SL_2(3)$ is the subgroup of $GL_2(3)$ leaving k fixed. The second statement is due to the fact that $\Delta^{1/3}$ generates an abelian extension of degree 3 over the cyclotomic field Q_{ab}, and the fact that

$$(SL_2(\mathbb{Z}_3) : E_3') = 3$$

in the preceding lemma.

It was convenient to visualize the theorem locally at 3, but in the applications, we use it in situations when 3 splits.

Theorem 4.2. *Let* L *be a set of primes not containing* 3.

(i) *If*
$$G_{3L} = GL_2(\mathbb{Z}_3) \times G_L$$
then
$$G_{k,3L} = E_3 \times G_L .$$

(ii) *If* $G_{3L} = G_3 \times G_L$ *and* $G_{k,3L} = E_3 \times G_L$, *then* $G'_{k,3L} = E'_3 \times G'_L$, *and*
$$K_{3L} \cap k_{ab,3L} = Q_{ab,3L}(\Delta^{1/3}) .$$

Proof. The first assertion follows from Theorem 4.1, since k is already contained in K_3. The second follows from the lemma, and the fact that
$$(G'_{3L} : G'_{k,3L}) = (G'_3 : E'_3) = 3 .$$

The lattice of fields illustrating the preceding discussion can be drawn as follows.

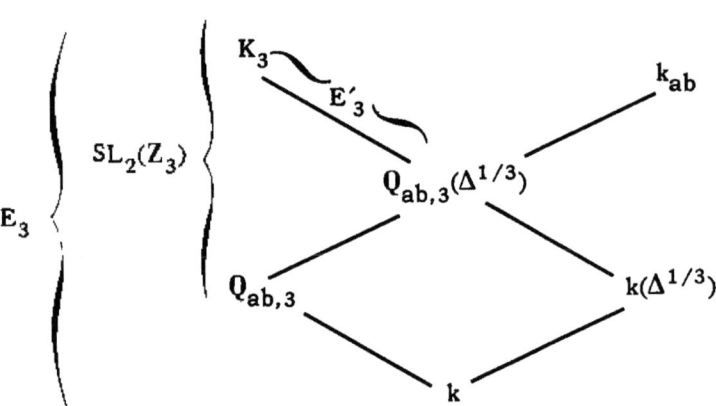

§5. $K \cap k_{ab}$ in other cases

Theorem 5.1. *Let* $k = Q(\sqrt{\Delta})$, $\Delta_0 = -q$ *where* q *is an odd prime. Assume that*
$$G_2 = GL_2(Z_2) \quad \text{and} \quad G = G_2 \times_\Delta G_{[2]}.$$
Then $K \cap k_{ab} = Q_{ab}F$, *where* F *is an abelian extension of* k, *and*
$$[F:k] = [Q_{ab}F : Q_{ab}].$$

Furthermore:

(i) *If* $q \neq 3$, *then* $F = k(\Delta^{1/4}, A_2)$ *and* $[F:k] = 6$.

(ii) *If* $q = 3$, *then* $F = k(A_2, \Delta^{1/4}, \Delta^{1/3})$ *and* $[F:k] = 18$.

Proof. The field diagram is as follows.

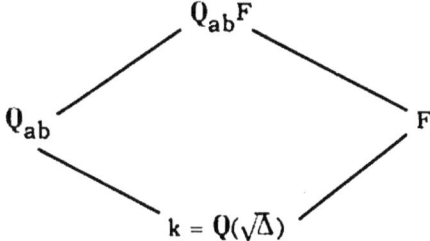

We note that k is contained in both K_2 and K_q. We have trivially $G_{2,k} = E_2$ and hence
$$G_k = E_2 \times G_{[2],k}.$$

By Theorem 3.1 and Lemma 4 (iii) of §2, we find
$$G'/G'_k = G'_2/G'_{k,2} = G'_2/E'_2 \approx G(4)'/E(4)',$$

and this last factor group has order 6. Since we have exhibited a cyclic extension of degree 6, namely F over k, whose intersection with Q_{ab} is obviously k, the theorem follows when $q \neq 3$.

Suppose that $q = 3$. Then we have to take into account the extra cyclic group of order 3 appearing in Theorem 3.1. Note that only 3 ramifies in $k(\Delta^{1/3})$. On the other hand, 2 must ramify in $Q(A_2)$ and 2 is unramified in $k = Q(\sqrt{-3})$. Hence 2 must ramify in $k(A_2)$. Hence $\Delta^{1/3}$ does not lie in $k(A_2)$, and it follows that

$$[k(A_2, \Delta^{1/4}, \Delta^{1/3}) : k] = 18 .$$

This combined with Theorem 3.1 immediately implies the case $q = 3$ of our theorem.

Theorem 5.2. *Let* $k = Q(i)$. *Assume that* $k \neq Q(\sqrt{\Delta})$. *Assume also that*

$$G = GL_2(Z_2) \times_\Delta G_{[2]}$$

is Serre's fibering. Then:

(i) *We have*

$$K \cap k_{ab} = Q_{ab} F \qquad \text{where} \qquad F = k(\Delta^{1/4}) .$$

Proof. We have

$$G_k = (\text{Ker } \chi_i) \times_\Delta G_{[2]} ,$$

whence by Lemma 1 of §3,

$$G'_k = (\text{Ker } \chi_i)' \times G'_{[2]}$$

$$G' = G'_2 \times G'_{[2]} .$$

Thus we have an inclusion of groups

$$SL_2(Z_2) \supset G'_2 \supset G'_k \supset (I+4M) \cap SL_2(Z_2) ,$$

with successive indices 2, 2, 12 as one sees from Theorem 2.5. This proves (i).

Since $\Delta \nmid -4Q^4$, we get $[F : k] = 4$, and $\Delta^{1/2}$ is contained in Q_{ab}. We know that $\Delta^{1/4}$ is contained in the field of 4-th division points, so the assertion (ii) is immediate from (i).

The lattice of fields is shown in the diagram.

```
                K ∩ k_ab = Q_ab(Δ^{1/4})
              ╱                        ╲
         Q_ab                            F = k(Δ^{1/4})
              ╲                        ╱
                    k(Δ^{1/2})
                        |
                        k
```

Remark. The hypothesis in both theorems that

$$G = GL_2(\mathbb{Z}_2) \times_\Delta G_{[2]}$$

is of course satisfied for Serre curves and for $X_0(11)$, whose Galois group was determined in Part I, §8.

$$k = \mathbb{Q}(\sqrt{-3})$$

§6. The action of $\mathcal{G}$ on $k(\Delta^{1/3})$

Eventually we want to mix the GL_2-extension with k_{ab}, and for this we have to make some properties of class field theory over k explicit on the cubic extension $k(\Delta^{1/3})$. We assume that $\Delta = \pm q^n$, where q is a prime > 3, and n is not divisible by 3. We let

$$k = \mathbb{Q}(\sqrt{-3}), \qquad F = k(\Delta^{1/3}) = k(q^{1/3}).$$

We let

$$\phi'' : \mathcal{G} \longrightarrow \mu_3 \approx \text{Gal}(F/k)$$

be the restriction homomorphism, composed with the identification of $\text{Gal}(F/k)$ with the group of cube roots of unity, through Kummer theory. We have to determine ϕ'' as explicitly as possible.

Case 1. $q \equiv \pm 1 \pmod 9$

In this case, F/k is ramified only at q, because $\Delta^{1/3}$ has one conjugate in $\mathbb{Z}_3$, i.e. $X^3 - \Delta$ has one root mod 9, so has one root mod 27 (because a cube mod 9 is also a cube mod 27), and we can refine such a root to a root of $X^3 - \Delta$ in $\mathbb{Z}_3$.

It follows that ϕ'' factors through

$$\mathcal{G} \longrightarrow \mathcal{G}_q = \mathfrak{o}_q^* / \mathfrak{o}^*.$$

In addition, since $q \neq 3$, it must also be that ϕ'' factors through

$$\mathcal{G} \longrightarrow \mathfrak{o}(q)^*/\mathfrak{o}^*.$$

Case 1a. $q \equiv -1 \pmod 9$

In this case, we must have $\left(\frac{k}{q}\right) = -1$. Then

$$\mathfrak{o}(q)^* = F_{q^2}^*$$

and we can factorize ϕ'',

$$\mathfrak{o}_q^* \longrightarrow \mathfrak{o}(q)^* \longrightarrow \mu_3.$$

The kernel of ϕ'' consists of the elements which are third powers, i.e.

$$\mathfrak{o}_q^{*3}.$$

Representatives for the cosets are given by $1, \omega, \omega^2$.

Case 1b. $q \equiv 1 \pmod{9}$

In this case, we must have $\left(\frac{k}{q}\right) = 1$, and we have a factorization

$$q = \pi\bar{\pi} \text{ in } \mathfrak{o}.$$

The extension F/k ramifies only over π and $\bar{\pi}$, and ϕ'' factors through

$$\mathfrak{o}(q)^* = \mathfrak{o}(\pi)^* \times \mathfrak{o}(\bar{\pi})^* \approx F_q^* \times F_q^*.$$

Let $\theta, \bar{\theta}$ be such that

$$\theta^3 = \pi \quad \text{and} \quad \bar{\theta}^3 = \bar{\pi}.$$

For each extension $k(\theta)$ and $k(\bar{\theta})$ we have the corresponding class field theory map

$$\phi''_\pi : \mathfrak{o}(\pi)^* \longrightarrow \text{Gal}(k(\theta)/k) \quad \text{and} \quad \phi''_{\bar{\pi}} : \mathfrak{o}(\bar{\pi})^* \longrightarrow \text{Gal}(k(\bar{\theta})/k).$$

Each of these Galois groups can be identified with μ_3, and

$$\phi''_q = \phi''_\pi \otimes \phi''_{\bar{\pi}}.$$

For $a \in \mathfrak{o}(\pi)^* \approx Z(q)^*$, let $\phi_\pi(a) = \zeta$. Then

$$\phi''_q(a^{-1}) = \phi''_\pi(a)\phi''_{\bar{\pi}}(a^{-1}) = \zeta\zeta^{-1} = 1.$$

Hence if $\beta \in \mathfrak{o}(q)^*$ and $N\beta = 1$ then $\phi''_q(\beta) = 1$. Consequently, ϕ''_q factors through the norm map,

$$\phi''_q : \mathfrak{o}_q^* \longrightarrow \mathfrak{o}(q)^* \longrightarrow Z(q)^* \longrightarrow \mu_3.$$

Since $Z(q)^*$ is cyclic, it has a unique subgroup of index 3, which consists of the cubes. The kernel

$$\mathfrak{o}_{q,1}^{*'}$$

consists of the inverse image of these cubes under the norm map, composed with reduction mod q. Equivalently, it is the inverse image of the cubes in Z_q^* under the norm map

$$\mathfrak{o}_q^* \xrightarrow{N} Z_q^*.$$

Case 2. $q \not\equiv \pm 1 \pmod 9$

In this case, both 3 and q are ramified in $F = k(q^{1/3})$, and hence the class field theory map ϕ'' factors through

$$\phi''_{3q} : (\mathfrak{o}_3^* \times \mathfrak{o}_q^*)/\mathfrak{o}^* \longrightarrow \mu_3.$$

Furthermore,

$$\phi''_{3q} = \phi''_3 \otimes \phi''_q.$$

The kernel of ϕ''_{3q} must then be determined through local class field theory, as the norm group locally.

Lemma 1. *Let* $\lambda = \sqrt{-3}$. *Then*

$$(1+3Z_3)(1+3\lambda Z_3) = 1 + 3\mathfrak{o}_3.$$

Proof. The set $1 + 3\lambda Z_3$ is not a group, but it is easily verified that the left hand side is a group, and it is also easily verified recursively that every element of $1 + 3\mathfrak{o}_3$ can be written as an element of the left hand side. We leave the details to the reader.

Lemma 2. $\quad (1+\lambda \mathfrak{o}_3)^3 = 1 + 3\lambda \mathfrak{o}_3.$

Proof. The left hand side is clearly contained in the right hand side. Conversely, suppose given an element of the form

$$1 + 3\lambda a, \qquad a \in \mathfrak{o}_3.$$

Then $1 + \lambda a$ is a cube root mod 9. Furthermore, given an element of the form

$$1 + 9b,$$

then $1 + 3b$ is a cube root mod 27. From here on, any standard refinement procedure takes hold to show that a cube root can be extracted, as desired.

Theorem 6.1. *Let* $q \not\equiv \pm 1 \pmod 9$. *Then*

$$\operatorname{Ker} \phi''_3 = \pm(1+3\mathfrak{o}_3).$$

Proof. We note that q, and hence q^2 is a local norm, and

$$q^2 \not\equiv 1 \pmod 9 .$$

Consequently the closed subgroup generated by q^2 is $1 + 3Z_3$, and is contained in the kernel of ϕ''_3. This kernel also contains all cubes, and therefore by Lemmas 1, 2 it contains $\pm(1+3\mathfrak{o}_3)$. Since $\pm(1+3\mathfrak{o}_3)$ has index 3 in $\mathfrak{o}_3^*$, the kernel is what we have stated.

Remark. The cosets of the kernel in Theorem 6.1 are the sets

$$\pm(\zeta + 3\mathfrak{o}_3)$$

where ζ ranges over the cube roots of unity. We shall denote them by

$$\mathfrak{o}_{3,\zeta}^* .$$

Observe that this is what we denote also by

$$r^{-1}(\zeta) ,$$

where r is reduction mod 3.

§7. The constant for Serre fiberings
$$k = \mathbb{Q}(\sqrt{-3}), \quad M = 2q, \quad q \text{ odd prime} \neq 3, \quad \Delta = \pm q^n$$

We assume the conditions stated in the title of the section, and also that

$$G_{6q} = S_{2q} \times GL_2(\mathbb{Z}_3),$$

where $S_{2q} = G_{2q}$ is Serre's subgroup. We determine the constant in Theorems 7.3, 7.4, 7.5. We first have to determine $\tilde{\mathcal{G}}_{6q}$, and decompose it as a union of products so that we can apply Part II, Theorem 6.1. We let

$$F = k(\Delta^{1/3}) = k(q^{1/3})$$

as before, and Δ has the stated value $\pm q^n$. From Theorem 4.2 we know that

$$K_{6q} \cap k_{6q}^{ab} = Q_{6q}^{ab}(\Delta^{1/3}).$$

We note that $\tilde{\mathcal{G}}_{6q}$ is of index 3 in

$$\tilde{\mathcal{G}}_{2q} \times (GL_2(\mathbb{Z}_3) \times_N \mathcal{C}_3).$$

The mixing is due to the two maps ϕ' and ϕ'',

$$G \longrightarrow GL_2(3) \xrightarrow{\phi'} \mu_3 \approx \text{Gal}(F/k)$$

$$\mathcal{C} \longrightarrow \mathcal{C}_{3q} \xrightarrow{\phi''} \mu_3.$$

Consequently

(1) $$\mathcal{G}_{6q} = G_{6q} \times_{N, \phi' = \phi''} \mathcal{C}_{6q} = G_{k, 6q} \times_{N, \phi' = \phi''} \mathcal{C}_{6q}.$$

To determine this fiber product, we write a decomposition for $G_{k, 6q}$ and $\mathcal{C}_{6q}$.

We abbreviate as before

$$E_3 = SL_2(\mathbb{Z}_3)(I + 3M_3).$$

Then

(2) $$G_{k,6q} = [(E_2 \times E_q) \cup (O_2 \times O_q)] \times E_3.$$

For each $\zeta \in \mu_3$ we let X_ζ be the set of elements $x \in X$ such that $\phi(x) = \zeta$, and $\phi = \phi'$ or $\phi = \phi''$ depending on the context. Thus for instance

$$E_{3,\zeta} = \{\sigma' \in E_3, \phi'(\sigma') = \zeta\}$$

$$\mathcal{A}_{q,\zeta} = \{\sigma'' \in \mathcal{A}_q, \phi''(\sigma'') = \zeta\}$$

and so forth. From (1) and (2) we get the decomposition

(3) $$\mathcal{G}_{6q} = \bigcup_{\zeta \in \mu_3} (E_2 \times E_{3,\zeta} \times E_q) \times_N (\mathcal{A}_2 \times (\mathcal{A}_3 \times \mathcal{A}_q)_\zeta)$$

$$\bigcup_{\zeta \in \mu_3} (O_2 \times E_{3,\zeta} \times O_q) \times_N (\mathcal{A}_2 \times (\mathcal{A}_3 \times \mathcal{A}_q)_\zeta).$$

We lift this to $\tilde{\mathcal{G}}$, to obtain

(4) $$\boxed{\begin{aligned}\tilde{\mathcal{G}}_{6q} &= \bigcup_\zeta [(E_2 \times E_{3,\zeta} \times E_q) \times_N (o_2^* \times (o_3^* \times o_q^*)_\zeta)] \\ &\quad \bigcup_\zeta [(O_2 \times E_{3,\zeta} \times O_q) \times_N (o_2^* \times (o_3^* \times o_q^*)_\zeta)].\end{aligned}}$$

The set $(o_3^* \times o_q^*)_\zeta$ is the inverse image of ζ under the map

$$\phi''_{3q} = \phi''_3 \otimes \phi''_q : o_3^* \times o_q^* \longrightarrow \mu_3.$$

We observe that the 2-factor in (4) splits off. Consequently we obtain:

(5) $$\mathrm{Num}_{6q}(\tilde{\mathcal{G}}) = \mathrm{Num}_2^+ \sum_\zeta \iint_{(o_3^* \times o_q^*)_\zeta} h_{E_{3,\zeta}}(\mathrm{Tr}\, z_3, Nz_3) h_{E_q}(\mathrm{Tr}\, z_q, Nz_q) dz_3\, dz_q$$

$$+ \mathrm{Num}_2^- \sum_\zeta \iint_{(o_3^* \times o_q^*)_\zeta} h_{E_{3,\zeta}}(\mathrm{Tr}\, z_3, Nz_3) h_{O_q}(\mathrm{Tr}\, z_q, Nz_q) dz_3\, dz_q.$$

To give the value of $\mathrm{Num}_{6q}(\tilde{\mathcal{G}})$ we have to distinguish cases depending on the congruence properties of q.

If $q \equiv \pm 1 \pmod 9$ then

$$(o_3^* \times o_q^*)_\zeta = o_3^* \times o_{q,\zeta}^*,$$

and consequently the 3-adic integral splits off. We obtain:

Theorem 7.1. *Let $\tilde{\mathcal{G}}_{6q}$ be the group associated with a Serre curve, and let k, 2q, Δ be as in the title of the section. Assume in addition that $q \equiv \pm 1 \pmod 9$. Then*

$$\mathrm{Num}_{6q}(\tilde{\mathcal{G}}) = \mathrm{Num}_2^+ \sum_\zeta \mathrm{Num}_3(E_{3,\zeta}, o_3^*) \mathrm{Num}_q(E_q, o_{q,\zeta}^*)$$

$$+ \mathrm{Num}_2^- \sum_\zeta \mathrm{Num}_3(E_{3,\zeta}, o_3^*) \mathrm{Num}_q(O_q, o_{q,\zeta}^*).$$

On the other hand, if $q \not\equiv \pm 1 \pmod 9$, then

$$\mathrm{Num}_{6q}(\tilde{\mathcal{G}}) = \mathrm{Num}_2^+ \sum_\zeta \sum_{\zeta_3 \zeta_q = \zeta} \mathrm{Num}_3(E_{3,\zeta}, o_{3,\zeta_3}^*) \mathrm{Num}_q(E_q, o_{q,\zeta_q}^*)$$

$$+ \mathrm{Num}_2^- \sum_\zeta \sum_{\zeta_3 \zeta_q = \zeta} \mathrm{Num}_3(E_{3,\zeta}, o_{3,\zeta_3}^*) \mathrm{Num}_q(O_q, o_{q,\zeta_q}^*).$$

The next section is devoted to evaluating the terms occurring in these sums.

In the following section, we shall establish a table of values for the numerator at the prime 3. We denote by ω, ω^2 the roots of unity in μ_3 which are $\neq 1$. The table is as follows.

		o_3^*		
		1	ω	ω^2
E_3	1	2/27	0	0
	ω	8/81	8/81	8/81
	ω^2	8/81	8/81	8/81

The table gives $\text{Num}_3(E_{3,\zeta}, o^*_\zeta\cdot)$. We note that these values are all equal to each other if $\zeta \neq 1$, and equal to $8/81$. Thus we find for instance

$$\text{Num}_3(E_{3,\zeta}, o^*_3) = 8/27 \quad \text{if} \quad \zeta \neq 1.$$

$$\text{Num}_3(E_{3,1}, o^*_3) = 2/27.$$

We shall also find:

Theorem 7.2. *For all values of ζ,*

$$\text{Num}_q(O_q, o^*_{q,\zeta}) = \frac{1}{6}(1+r)(1-r)^2$$

is independent of ζ, where $r = 1/q$.

This allows us to make simplifications in the sums of Theorem 7.1. Start with the case $q \equiv \pm 1 \pmod{9}$. We separate the first sum over ζ for $\zeta = 1$ and $\zeta = \omega, \omega^2$. In the second sum, we can replace the q-terms by their constant value, and then the sum over ζ for the terms $\text{Num}_3(E_{3,\zeta}, o^*_3)$ give $\text{Num}_3(E_3, o^*_3)$. We get:

$$\text{Num}_{6q}(\tilde{\mathcal{G}}_{6q}) = \text{Num}^+_2 \left[\frac{2}{27} \text{Num}_q(E_q, o^*_{q,1}) + \frac{16}{27} \text{Num}_q(E_q, o^*_{q,\omega}) \right]$$
$$+ \frac{1}{3} \text{Num}^-_2 \text{Num}^+_3 \text{Num}^-_q.$$

Since we are in the ramified case, we know from the tables in Part II, §9 and §10 that

$$\text{Num}^+_3 = \text{Num}_3 = \frac{2}{3}.$$

Consequently the above expression can be rewritten in a more generic form as follows.

Theorem 7.3. *Suppose that $q \equiv \pm 1 \pmod 9$. Then*

$$\text{Num}_{6q}(\tilde{\mathcal{G}}_{6q}) = \frac{1}{3} \text{Num}_3 \left[\text{Num}^+_2 \left(\frac{1}{3} \text{Num}_q(E_q, o^*_{q,1}) + \frac{8}{3} \text{Num}_q(E_q, o^*_{q,\omega}) \right) + \text{Num}^-_2 \text{Num}^-_q \right].$$

Consider the other case $q \not\equiv \pm 1 \pmod 9$. Again in the first sum we take $\zeta = 1$ and $\zeta = \omega, \omega^2$ separately. We see that

$$\sum_{\zeta = \omega, \omega^2} \sum_{\zeta_3 \zeta_q = \zeta} \text{Num}_3(E_{3,\zeta}, o^*_{3,\zeta_3}) \text{Num}_q(E_q, o^*_{q,\zeta_q})$$

is equal to

$$\sum_{\zeta=\omega,\omega^2} \sum_{\zeta_3\zeta_q=\zeta} \tfrac{8}{81} \text{Num}_q(E_q, o^*_{q,\zeta_q}) = \tfrac{2\cdot 8}{81} \text{Num}^+_q$$

because the sum over ζ_3 becomes trivial.

On the other hand, we can use Theorem 7.2 for the second sum, and then the terms summed over ζ_3 add up to $\text{Num}_3(E_{3,\zeta}, o^*_3)$. We can then sum over ζ. We find:

$$\text{Num}_{6q}(\tilde{\mathcal{G}}) = \text{Num}^+_2 \left[\tfrac{2}{27} \text{Num}_q(E_q, o^*_{q,1}) + \tfrac{16}{81} \text{Num}^+_q \right]$$
$$+ \tfrac{1}{3} \text{Num}^-_2 \text{Num}^+_3 \text{Num}^-_q .$$

Factoring out in the same manner as in the preceding case, we obtain:

Theorem 7.4. *Assume that* $q \not\equiv \pm 1 \pmod 9$. *Then*

$$\text{Num}_{6q}(\tilde{\mathcal{G}}_{6q}) = \tfrac{1}{3} \text{Num}_3 \left[\text{Num}^+_2 \left(\tfrac{1}{3} \text{Num}_q(E_q, o^*_{q,1}) + \tfrac{8}{9} \text{Num}^+_q \right) + \text{Num}^-_2 \text{Num}^-_q \right] .$$

Remark. The expressions of Theorem 7.4 and 7.3 are identical, except for the terms

$$\tfrac{8}{9} \text{Num}^+_q \quad \text{and} \quad \tfrac{8}{3} \text{Num}_q(E_q, o^*_{q,\omega}) .$$

However, these two terms are close together for large values of q.

We still have to give the values for the numerators involving E_q.

Theorem 7.5. *Suppose* $\left(\tfrac{k}{q} \right) = 1$. *Then for all* ζ,

$$\text{Num}_q(E_q, o^*_{q,\zeta}) = \tfrac{1}{6}(1+r)(1-r)^2 = \tfrac{1}{3} \text{Num}^+_q .$$

Hence in this case,

$$\text{Num}_{6q}(\tilde{\mathcal{G}}_{6q}) = \tfrac{1}{3} \text{Num}_3 \text{Num}_{2q}(\mathcal{S}_{2q}) .$$

where $\mathcal{S}_{2q}$ *is Serre's group of Part II, §10.*

Proof. Obvious from Theorems 7.3, 7.4, and Part II, Theorem 10.2.

Corollary. If $\left(\frac{k}{q}\right) = 1$, then

$$\text{Den}_{6q}(\tilde{\mathcal{G}}_{6q}) = \tfrac{1}{3} \text{Den}_3 \, \text{Den}_{2q}(\mathcal{S}_{2q})$$

and

$$C_{6q}(\tilde{\mathcal{G}}_{6q}) = C_3 C_{2q}(\mathcal{S}_{2q}).$$

Proof. The value for the denominator follows from Theorem 6.4 of Part II, because $\tilde{\mathcal{G}}_{6q}$ is obtained from a fibering of degree 3 over the Serre group. The value for the constant itself is then obvious.

Theorem 7.6. Suppose $\left(\frac{k}{q}\right) = -1$. Then

$$\text{Num}_q(E_q, \mathfrak{o}_{q,\zeta}^*) = \tfrac{1}{6}(1+r)(1-r)^2 - 2r \int_{\mathfrak{o}_{q,\zeta}^*} r^{v(z)} \, dz \, .$$

Proof. Given in the next section, formula (4).

The values for the integral are given in Lemmas 1 and 2 at the end of the next section.

§8. Computation of integrals

We compute the appropriate integrals to justify the values given in the preceding section.

Computations at 3

We have $k = \mathbb{Q}(\sqrt{-3})$. Let $\lambda = 1 - \omega$. We note that

$$\mathfrak{o}^*(3) = \{\pm 1, \pm\omega, \pm\omega^2\} = \{\pm 1, \pm(1+\lambda), \pm(1-\lambda)\}.$$

On the other hand, we had already seen in §4:

$$E_{3,1}(3) = \begin{cases} I \text{ and } -I \\ 6 \text{ matrices with trace } 0 \mod 3 \end{cases}$$

and for $\zeta \neq 1$:

$$E_{3,\zeta}(3) = \begin{cases} 8 \text{ matrices which are non-scalar} \mod 3 \\ 4 \text{ have trace } 1 \mod 3 \\ 4 \text{ have trace } -1 \mod 3. \end{cases}$$

Observe that all elements of $\mathfrak{o}^*(3)$ have trace $\not\equiv 0 \mod 3$. This gives rise to several orthogonalities in the integrals we have to evaluate.

From the description of $E_{3,\zeta}$ for $\zeta \neq 1$, and Part II, Theorem 7.1 we find:

$$h_{E_{3,\zeta}}(t,s) = \begin{cases} 4r^2 & \text{if } t \equiv 0 \text{ and } s \equiv 1 \mod 3 \\ 0 & \text{otherwise}. \end{cases}$$

Therefore for $\omega \neq 1$, we get

$$\text{Num}_3(E_{3,\omega}, \mathfrak{o}^*_{3,\zeta}) = \tfrac{4}{9} \mu(\mathfrak{o}^*_{3,\zeta}) = 8/81,$$

and the same value with ω replaced by ω^2.

Since the traces of $\mathfrak{o}^*_3$ don't match the traces of matrices with trace 0 in $E_{3,1}$, we get

$$\text{Num}_3(E_{3,1}, o^*_{3,1}) = \int_{o^*_{3,1}} [h_{I+3M} + h_{-I+3M}(\text{Tr } z, Nz)] dz .$$

The integral of h_{I+3M} over $o^*_{3,1}$ is the same as its integral over $1 + 3o_3$. The integral of h_{-I+3M} over $o^*_{3,1}$ is the same as its integral over $-1 + 3o_3$. Hence their sum is equal to

$$2 \int_{3o_3} h_{3M}(\text{Tr } z, Nz) dz = 2/27$$

by Part II, §8, Lemma 6.

Next, let $\lambda = 1 - \omega$, so λ is the prime in o. We get:

$$\text{Num}_3(E_{3,1}, o^*_{3,\omega}) = \int_{\pm(1+\lambda) + 3o_3} h_{I+3M} + \int_{\pm(1+\lambda) + 3o_3} h_{-I+3M}$$

$$= \int_{1+\lambda+3o_3} h_{I+3M} + \int_{-(1+\lambda)+3o_3} h_{-I+3M}$$

$$= \int_{\lambda+3o_3} h_{3M} + \int_{-\lambda+3o_3} h_{3M}$$

$$= 0 .$$

(See Lemmas 4, 5 of Part II, §8.)

This concludes the proof of the evaluation of the entries of the table of 3-values.

Computations at q

We first deal with the terms involving O_q because they come out more simply and uniformly. Note that all elements of O_q are non-scalar mod q because the scalars have square determinant mod q.

Since q is an odd prime $\neq 3$, it follows that q is unramified in k.

In Part II, §10, Lemma 1, we had found:

$$h_{O_q}(\text{Tr } z, Nz) = \begin{cases} 0 & \text{unless } Nz \text{ is a non-square unit and otherwise:} \\ 1+r & \text{if } \left(\frac{k}{q}\right) = 1 \\ 1-r & \text{if } \left(\frac{k}{q}\right) = -1 \end{cases}$$

On the other hand, for any ζ, under the norm map,

$$N : o_{q,\zeta}^* \longrightarrow Z_q^*$$

the inverse images N^{-1}(squares) and N^{-1}(non squares) differ by a multiplicative translation. Indeed, there exists an element $b \in o_q^*$ such that Nb is not a square (because q is unramified). Hence Nb^3 is not a square, and $b^3 \in o_{q,1}^*$. Multiplication by b^3 on $o_{q,\zeta}^*$ permutes the subsets whose norms are squares and non-squares respectively.

In particular, the measures of the sets of elements in $o_{q,\zeta}^*$ whose norms are squares or non-squares respectively, are equal, and in fact equal to

$$\tfrac{1}{2}\mu(o_{q,\zeta}^*).$$

We now find:

$$\int_{o_{q,\zeta}^*} h_{O_q}(\text{Tr } z, Nz)\, dz = \tfrac{1}{2}\mu(o_{q,\zeta}^*) \cdot \begin{cases} 1+r & \text{if } \left(\frac{k}{q}\right) = 1 \\ 1-r & \text{if } \left(\frac{k}{q}\right) = -1 \end{cases}$$

(1)
$$= \tfrac{1}{6}(1-r)^2(1+r)$$

as stated in Theorem 7.2.

We shift to E_q and M. In case $\left(\frac{k}{q}\right) = 1$, the factor $\psi_q(k) = 0$ in Theorem 8.2 of Part II. Hence we get

(2)
$$\int_{o_{q,\zeta}^*} h_M(\text{Tr } z, Nz)\, dz = (1+r)\mu(o_{q,\zeta}^*) = \tfrac{1}{3}(1+r)(1-r)^2.$$

Consequently by subtraction,

(3)
$$\int_{o_{q,\zeta}^*} h_{E_q}(\text{Tr } z, Nz)\, dz = \tfrac{1}{6}(1-r)^2(1+r).$$

For the rest of this section, we assume

$$\left(\frac{k}{q}\right) = -1.$$

In this case, $\psi_q(k) = 2r$. Using Theorem 8.1 of Part II, we find

(4) $$\int_{o_{q,\zeta}^*} h_M(\operatorname{Tr} z, Nz)\, dz = \tfrac{1}{3}(1+r)(1-r)^2 - 2r \int_{o_{q,\zeta}^*} r^{v(z)}\, dz.$$

Subtracting (1) gives the integral with M replaced by E_q. By Part II, Theorem 8.3, we get:

$$\int_{o_{q,\zeta}^*} r^{v(z)}\, dz = \tfrac{r^3}{1+r} |Z(q)^* \cap o(q)_\zeta^*|$$
$$+ r^2 |o(q)_\zeta^* - Z(q)^*|.$$

We deal first with $\zeta = 1$. Then $o_{q,1}^*$ consists of the cubes in o_q^*. Since $\left(\frac{k}{q}\right) = -1$, all elements of $Z(q)^*$ are cubes, and $Z(q)^*$ is contained in $o(q)_1^*$. Hence

$$|o(q)_1^* \cap Z(q)^*| = q-1$$

$$|o(q)_1^* - Z(q)^*| = \tfrac{q^2-1}{3} - (q-1).$$

This gives

Lemma 1. *Assume* $\left(\frac{k}{q}\right) = -1$. *Then*

$$\int_{o_{q,1}^*} r^{v(z)}\, dz = \tfrac{r^2(1-r)}{1+r} + \tfrac{1-r^2}{3} - r + r^2$$
$$= \tfrac{(1-r)(1-r+r^2)}{3(1+r)}.$$

Lemma 2. *Assume* $\left(\frac{k}{q}\right) = -1$. *If* $\zeta \neq 1$ *then*

$$\int_{o_{q,\zeta}^*} r^{v(z)}\, dz = \tfrac{1}{3}(1-r^2).$$

Proof. In this case, $o(q)^*_\zeta \cap Z(q)^*$ is empty, so the integral is equal to

$$r^2 |o(q)^*_\zeta| = \mu(o^*_{q,\zeta}) = \frac{1}{3} \mu(o^*_q),$$

which gives the desired value.

$$k = \mathbf{Q}(i)$$

§9. The constant for Serre fiberings, q odd $\neq 3$

Throughout this section we assume that $k = \mathbb{Q}(i)$, and that

$$\Delta = -qc^4, \qquad \Delta_0 = -q,$$

where $c \in \mathbb{Z}$ and q is an odd prime. We let

$$F = k(\Delta^{1/4}),$$

so that by Theorem 5.2, F is cyclic over k of degree 4. We let B be the Galois group, $B = \text{Gal}(F/k)$, identified by Kummer theory with the group $\{\pm 1, \pm i\}$. We assume that

$$G_{2q} = GL_2(\mathbb{Z}_2) \times_\Delta GL_2(\mathbb{Z}_q).$$

The purpose of this section is to determine the constant

$$\text{Num}_{2q}(\tilde{\mathcal{G}}) = \text{Num}_{2q}(\tilde{\mathcal{G}}_{2q}).$$

This amounts to finding a decomposition of $\tilde{\mathcal{G}}_{2q}$, and computing integrals. We note that $k(\sqrt{\Delta})$ is unramified over k at 2, so the inertia group at 2 is the subgroup $\{\pm 1\}$ of B. From local class field theory, we have two local maps

$$\phi''_2 : \mathfrak{o}_2^* / N_2 \longrightarrow B \qquad \text{and} \qquad \phi''_q : \mathfrak{o}_q^* / N_q \longrightarrow B.$$

The image of ϕ''_2 is $\{\pm 1\}$. We put elements ζ of B as indices to indicate the inverse image of ζ under the maps ϕ''_2, ϕ''_q and the combined map

$$\phi'' = \phi''_2 \otimes \phi''_q : \mathfrak{o}_{2q}^* \longrightarrow B.$$

Then by definition,

(1) $$\mathfrak{o}_{2q,\zeta}^* = (\mathfrak{o}_{2,1}^* \times \mathfrak{o}_{q,\zeta}^*) \cup (\mathfrak{o}_{2,-1}^* \times \mathfrak{o}_{q,-\zeta}^*)$$

and

$$\tilde{\mathcal{G}}_{2q} = \bigcup_\zeta \mathfrak{o}_{2q,\zeta}^*.$$

Furthermore,
$$o^*_{q,-\zeta} = - o^*_{q,\zeta}.$$

So much for the class field theory side, we don't need to know any more about the maps ϕ''. On the matrix side, we let
$$G_{k,2} = V = \text{Ker } \chi_i \quad \text{in} \quad GL_2(\mathbb{Z}_2).$$

We have $G_2 = GL_2(\mathbb{Z}_2)$, $G_q = GL_2(\mathbb{Z}_q)$ and
$$G_{k,2q} = V \times_\Delta G_q.$$

We have homomorphisms
$$\phi'_2 : G_{k,2} \longrightarrow B \quad \text{and} \quad \phi'_q : G_q \longrightarrow B,$$

and by definition, for $\zeta \in B$, the inverse image under $\phi'_{2q} = \phi'_2 \otimes \phi'_q$ is
$$G_{k,2q,\zeta} = V_\zeta \times G_{q,\zeta^2}.$$

Then

(2) $$G_{k,2q} = \bigcup_\zeta (V_\zeta \times G_{q,\zeta^2}).$$

Putting (1) and (2) together, we obtain the decomposition

(3)
$$\tilde{\mathcal{G}}_{2q} = \bigcup_\zeta \left[\left(V_\zeta \times_N o^*_{2,1}\right) \times \left(G_{q,\zeta^2} \times_N o^*_{q,\zeta}\right) \right]$$
$$\bigcup_\zeta \left[\left(V_\zeta \times_N o^*_{2,-1}\right) \times \left(G_{q,\zeta^2} \times_N o^*_{q,-\zeta}\right) \right].$$

By definition of even and odd elements (for an odd prime q, the elements in G_q whose determinant is a square or not), we see that
$$G_{q,1} = E_q \quad \text{and} \quad G_{q,-1} = O_q.$$

We also have $-E_q = E_q$ and $-O_q = O_q$. Making the change of variables $z \mapsto -z$ in the integral, we see that if $X_q = E_q$ or O_q,

$$\text{Num}_q(X_q, o_{q,\zeta}^*) = \text{Num}_q(X_q, o_{q,-\zeta}^*).$$

Since $o_{2,1}^* \cup o_{2,-1}^* = o_2^*$, we get the formula

(4) $$\text{Num}_{2q}(\tilde{\mathcal{G}}_{2q}) = \sum_{\zeta=1,-1} \text{Num}_2(V_\zeta, o_2^*)\text{Num}_q(E_q, o_{q,1}^*)$$
$$+ \sum_{\zeta=i,-i} \text{Num}_2(V_\zeta, o_2^*)\text{Num}_q(O_q, o_{q,i}^*).$$

Observe also the further symmetry

$$\text{Num}_2(V_\zeta, o_2^*) = \text{Num}_2(V_{-\zeta}, o_2^*)$$

which comes from the change of variables $z \mapsto -z$ and

$$-V_\zeta = V_{-\zeta}.$$

Theorem 9.1. *Let $\mathcal{S}_{2q}$ be Serre's group. Then*

$$\text{Num}_{2q}(\tilde{\mathcal{G}}_{2q}) = \tfrac{1}{2} \text{Num}_{2q}(\mathcal{S}_{2q}).$$

Proof. Serre's group was defined in Part II, §10, as

$$\mathcal{S}_{2q} = (E_2 \times_N o_2^*) \times (E_q \times_N o_q^*) \cup (O_2 \times_N o_2^*) \times (O_q \times_N o_q^*).$$

Note that

$$V \cap E_2 = V_1 \cup V_{-1} \qquad \text{and} \qquad V \cap O_2 = V_i \cup V_{-i}.$$

In $E_q \times_N o_q^*$ we may replace o_q^* by $o_{q,1}^* \cup o_{q,-1}^*$ without changing the value, because $E_q \times_N o_{q,i}^* = E_q \times_N o_{q,-i}^* = 0$. Similarly, in $O_q \times_N o_q^*$ we may replace o_q^* by $o_{q,i}^* \cup o_{q,-i}^*$. Then we obtain

$$\text{Num}_{2q}(\mathcal{S}_{2q}) = \sum_{\zeta=1,-1}\sum_{\zeta'=1,-1} \text{Num}_2(V_\zeta, o_2^*)\text{Num}_q(E_q, o_{q,\zeta'}^*)$$
$$+ \sum_{\zeta=i,-i}\sum_{\zeta'=i,-i} \text{Num}_2(V_\zeta, o_2^*)\text{Num}_q(O_q, o_{q,\zeta'}^*).$$

Each term is invariant under $\zeta \mapsto -\zeta$ or $\zeta' \mapsto -\zeta'$. All the terms are equal within each double sum, and $\mathrm{Num}_{2q}(\tilde{\mathcal{G}}_{2q})$ consists of those terms for which $\zeta = \zeta'$. This proves the theorem.

Theorem 9.2. *We have*

$$\mathrm{Den}_{2q}(\tilde{\mathcal{G}}_{2q}) = \tfrac{1}{2}\,\mathrm{Den}_{2q}(\mathcal{S}_{2q})$$

and hence for the constant, the same value as in Part II, §10,

$$C_{2q}(\tilde{\mathcal{G}}_{2q}) = C_{2q}(\mathcal{S}_{2q}).$$

Proof. The group $\tilde{\mathcal{G}}_{2q}$ is obtained by a fibration which is cyclic of degree 2 over the field of roots of unity. The first assertion concerning the denominator follows from Part II, Theorem 6.4. The second is then obvious by the preceding theorem.

$$k = \mathbf{Q}(\sqrt{\Delta})$$

§10. The action of $\mathcal{G}$ on $k(A_2, \Delta^{1/4})$ when $k = Q(\sqrt{\Delta})$

Throughout this section, we assume that $\Delta = \Delta_0 c^4$, $c \in Z$, *and* $\Delta_0 = -q$, *where* q *is an odd prime,*
$$-q \equiv 5 \pmod 8.$$

[In the applications, $q = 11$ *or* 43.*] We let* $k = Q(\sqrt{\Delta})$.

We are interested in the class field theory map
$$\phi_{2q} : o_{2q}^* = o_2^* \times o_q^* \longrightarrow \text{Gal}(k(A_2, \Delta^{1/4})/k).$$

We begin with the kernel of the map
$$\phi_q : o_q^* \longrightarrow \text{Gal}(k(A_2, \Delta^{1/4})/k).$$

Theorem 10.1. *Let* k_q *be the completion of* k *at* q. *Then the norm group in* o_q^* *from* $k_q(\Delta^{1/4})$ *is the group of squares, and* -1 *is not a square, so not a norm.*

Proof. Since q ramifies in k, we see that
$$o(q)^* \approx Z(q)^* \times Z(q),$$
and $o(q)^*$ is cyclic of order $q(q-1)$, so has a unique subgroup of index 2, which consists of the squares. The same is therefore true of o_q^*, as desired. It is obvious that -1 is not a square.

In the cases of interest to us for the completion of our tables, we have a special fact.

Theorem 10.2. *Let the elliptic curve be either the curve*
$$y^2 + y = x^3 + x^2,$$
or $X_0(11)$. *Then* $q = 43$ *in the first case, and* 11 *in the second. In both*

cases, we have
$$k_q(A_2) = k_q,$$
i.e. the prime above q in k splits completely in $k(A_2)$.

Proof. A translation in y puts the curve in the form $Y^2 = g(x)$, where

$$g(x) = x^3 + x^2 + 1/4 \quad \text{and} \quad g(x) = x^3 - x^2 - 10x - 20 + 1/4$$

respectively. In both cases, the derivative $g'(x)$ has no multiple roots mod q, and q ramifies in k. Hence there is a factorization

$$g(x) \equiv g_1(x) g_2(x)^2 \pmod{q}$$

where $g_1(x)$, $g_2(x)$ are linear. Since q ramifies in k, the existence of a prime of degree 1 implies the assertion of the theorem.

In particular, for the special case of Theorem 10.2 we find that

$$k_q(A_2, \Delta^{1/4}) = k_q(\Delta^{1/4}).$$

Next we work at the prime 2. We do not need to determine the kernel of the local map

$$\phi_2 : \mathfrak{o}_2^* \longrightarrow \text{Gal}(k(A_2, \Delta^{1/4})/k)$$

explicitly. The congruence $-q \equiv 5 \pmod{8}$ implies that 2 remains prime in k.

Theorem 10.3. *Let* $F = k(A_2, \beta)$ *where* $\beta^4 = -q$. *Then*

$$[Fk_2 : k_2] = 6,$$

and Fk_2 *is totally ramified over* k_2. *The norm group* N_2 *in* $\mathfrak{o}_2^*$ *has index* 6. *Its cosets are represented by*

$$\pm 1, \quad \pm \omega, \quad \pm \omega^2,$$

where $\omega^3 = 1$ *and* $\omega \neq 1$.

Proof. Since 2 remains prime in k and ramifies in $Q(A_2)$, it follows that 2 ramifies in $k(A_2)$ over k. Hence $k_2(A_2)$ has degree 3 over k_2 and is totally ramified of order 3. On the other hand, if we put

$$\lambda = \beta - 1,$$

then it is easy to see that λ satisfies an equation

$$\lambda^2 + 2\lambda - 2\eta = 0$$

where η is defined by

$$\sqrt{-q} = 1 + 2\eta.$$

Then η is a root of the equation

$$\eta^2 + \eta + m = 0, \qquad \text{where} \qquad -q = 1 - 4m,$$

m is odd, and so η is a unit at 2, and the equation for λ is an Eisenstein equation, which shows that F is also ramified of order 2, whence totally ramified at 2.

All the elements of $1 + 2o_2$ are cubes, and hence the cube roots of unity represent o_2^* modulo cubes.

Finally, we have to see that -1 is not a local norm at 2. Let

$$(-1, 1, 1, \cdots)$$

be the idele which has component -1 at 2 and component 1 at all other primes of k. Then its Artin symbol is the same as the idele

$$(1, -1, -1, -1, \cdots).$$

The only ramified primes in F are 2 and q. We have already seen in Theorem 11.1 that -1 is not a local norm at q. Hence the Artin symbol of this idele is not trivial. This proves that -1 is not a local norm at 2, and concludes the proof of the theorem.

Remark. It can be shown that the norm group N_2 is generated by $1 + 4o_2$ and $-1 + 2\eta$. We do not need this in the sequel.

§11. The action of matrices on $k(A_4)$

We start with an elliptic curve over a field, defined by the equation

$$y^2 = f(x) = x^3 + c_4 x + c_6 .$$

We let e_1, e_2, e_3 be the roots of $f(x)$. We let

$$\delta = 4(e_1 - e_2)(e_2 - e_3)(e_3 - e_1), \qquad \delta^2 = \Delta .$$

If (x_i, y_i) ($i = 1,2,3$) are points on the curve, then we have the addition formula

$$x_1 + x_2 + x_3 = \left(\frac{y_1 - y_2}{x_1 - x_2}\right)^2$$

whenever

$$(x_1, y_1) + (x_2, y_2) + (x_3, y_3) = 0 ,$$

the addition being addition on the curve, and 0 the origin, i.e. the point at infinity. We then also have

$$\frac{y_1 - y_2}{x_1 - x_2} = \frac{y_2 - y_3}{x_2 - x_3} = \frac{y_3 - y_1}{x_3 - x_1} .$$

The points of order 2 are given in terms of coordinates by

$$(e_i, 0), \quad i = 1, 2, 3 .$$

The Weierstrass form yields

$$e_1 + e_2 + e_3 = 0 .$$

We need the coordinates for points of order 4 which divide these three points of order 2. If $P = (x, y)$ and $2P = (e_i, 0)$, then the x-coordinate of P is of the form

$$x = e_i \pm u_i \quad \text{where} \quad u_i^2 = 3e_i^2 + \sum_{j \neq j'} e_j e_{j'} .$$

and the four points P such that $2P = (e_i, 0)$ are of the form

$$(e_i + u_i, \pm v_i) \quad \text{where} \quad v_i^2 = u_i^2 (3e_i + 2u_i)$$

$$(e_i - u_i, \pm v_i') \quad \text{where} \quad v_i'^2 = u_i^2 (3e_i - 2u_i).$$

We fix a choice of u_1, v_1, u_2, v_2 and label

$$P_1 = (e_1 + u_1, v_1), \qquad P_2 = (e_2 + u_2, v_2).$$

We then define u_3, v_3 by the formula

$$P_3 = (e_3 + u_3, v_3) = -P_1 - P_2$$

so that

$$P_1 + P_2 + P_3 = 0.$$

Finally, we define v_i' to be those elements such that

$$(e_i + u_i, v_i) + (e_{i+1}, 0) + (e_i - u_i, v_i') = 0.$$

We observe in passing that

$$16 (u_1 u_2 u_3)^2 = -\delta^2,$$

so that we can express $\sqrt{-1}$ explicitly in the field of 4-division points, but we won't need this. We define

(1) $$w_i = \frac{v_i - v_i'}{(e_i + u_i) - (e_i - u_i)} = \frac{v_i - v_i'}{2u_i}.$$

We define

$$W = w_1 w_2 w_3.$$

The addition formula implies that

(2) $$w_i^2 = e_i - e_{i-1},$$

because

$$w_i^2 = e_i + u_i + e_{i+1} + (e_i - u_i) = 2e_i + e_{i+1} = e_i - e_{i-1}.$$

Therefore

(3) $$4W^2 = -\delta,$$

and we have obtained a root $\Delta^{1/4}$ in the field of 4-division points explicitly. We let

$$\tau = \begin{pmatrix} -1 & 2 \\ 2 & 1 \end{pmatrix} \quad \text{and} \quad \gamma = \begin{pmatrix} -1 & -1 \\ 1 & 0 \end{pmatrix}.$$

Theorem 11.1. *Let* $F = k(A_2, \beta)$, $B = \text{Gal}(F/k)$.

(i) *For one of the choices of $\sqrt{-q}$ defining η and β, the matrix τ changes β to $-\beta$. For the other choice, the matrix τ leaves β fixed.*

(ii) *The effect of γ on F generates the cyclic subgroup of order 3 in B.*

Proof. Since the determinant of τ is not $\equiv 1 \pmod{4}$, it follows that τ changes i to $-i$. Hence if τ leaves some fourth root of Δ fixed, it has to have a non-trivial effect on the other fourth root obtained by multiplication with i. This proves (i). As for (ii), since γ has period 3 and acts non-trivially on A_2, the assertion is obvious.

From now on, we assume that β is chosen so that $\tau\beta = \beta$. Then

$$E_1(4) = E'(4) \cup E'(4)\tau.$$

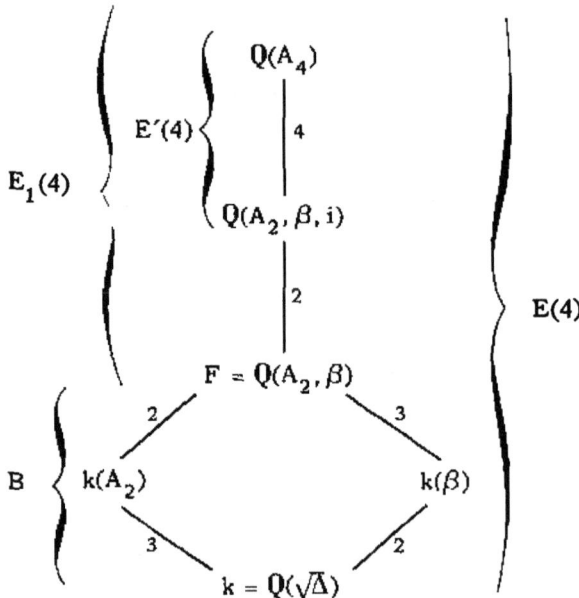

§12. Computation of integrals and the constant

Throughout this section, we work under the same conditions as §10. We assume that $\Delta = \Delta_0 c^4$, $c \in \mathbb{Z}$, *and* $\Delta_0 = -q$, *where* q *is an odd prime,*

$$-q \equiv 5 \pmod{8}.$$

We let $k = \mathbb{Q}(\sqrt{\Delta})$, *and* $\beta^4 = -q$. *We let* $F = k(A_2, \beta)$. *We let* $B = \mathrm{Gal}\,(F/k)$. *We assume that* $G_{2q} = S_{2q}$ *is Serre's fibering.*

We have the field diagram:

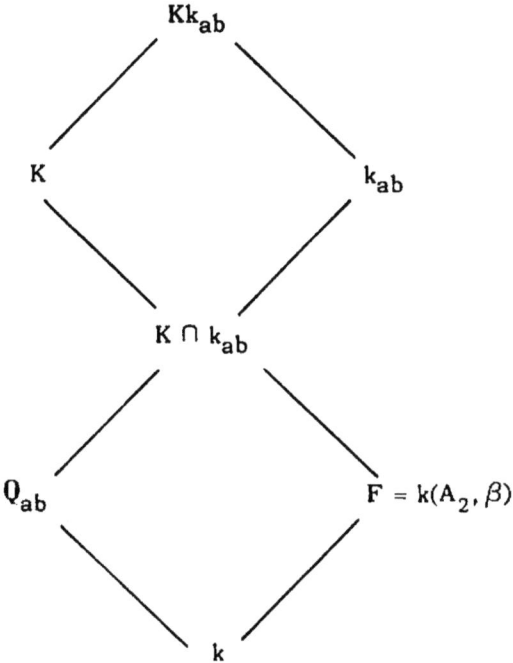

We know from Theorem 5.1 that $Q_{ab}F = K \cap k_{ab}$. We have two maps

$$\phi'_{2q} : S_{2q} \longrightarrow B$$

$$\phi''_{2q} : \tilde{\mathfrak{a}}_{2q} \approx \mathfrak{o}^*_{2q} \longrightarrow B$$

giving the Galois action on F for the GL_2-extension and for k_{ab}. We put

$$E_{2q} = E_2 \times E_q$$

as usual, where E_2 is the group of even elements, and E_q is the unique subgroup of index 2 in $GL_2(Z_q)$. Then
$$G_{k,2q} = E_{2q}.$$

We find
$$\tilde{\mathcal{G}}_{2q} = \{(\sigma_{2q}, a_{2q}) \in E_{2q} \times_N o^*_{2q}, \phi'_{2q}(\sigma_{2q}) = \phi''_{2q}(a_{2q})\}$$

or in other words,

(1) $$\tilde{\mathcal{G}}_{2q} = \bigcup_{\zeta \in B} E_{2q,\zeta} \times_N o^*_{2q,\zeta}$$

where we denote elements of B by ζ, and index by ζ the elements of E_{2q} and o^*_{2q} which lie in the inverse image of ζ by the corresponding map ϕ' or ϕ''.

We sometimes write -1 also for the unique element of B which has period 2. This is unambiguous, since B is cyclic. Thus if $\zeta \in B$, then $-\zeta$ is equal to ζ times this unique element.

Since $F \subset k(A_4)$, and $k(A_4) \cap K_q = k$ (because G_{2q} is the Serre fibering) we see that E_q acts trivially on F, whence

$$E_{2q,\zeta} = E_{2,\zeta} \times E_q.$$

Since -1 is not a square in o^*_q, we see that

$$o^*_{q,-1} = \text{non-squares in } o^*_q.$$

We obtain:

$$o^*_{2q,\zeta} = (o^*_{2,\zeta} \times o^*_{q,1}) \cup (o^*_{2,-\zeta} \times o^*_{q,-1}),$$

and therefore we obtain the decomposition:

(2) $$\boxed{\begin{aligned} E_{2q,\zeta} \times_N o^*_{2q,\zeta} &= \\ (E_{2,\zeta} \times_N o^*_{2,\zeta}) &\times (E_q \times_N o^*_{q,1}) \\ \cup\, (E_{2,\zeta} \times_N o^*_{2,-\zeta}) &\times (E_q \times_N o^*_{q,-1}). \end{aligned}}$$

Remark. We have

$$-E_{q,\zeta} = E_{q,-\zeta} \quad \text{and} \quad o^*_{2,-\zeta} = -o^*_{2,\zeta}.$$

The first equality results from the fact that $-I$ induces -1 on F, and the second likewise, in view of the fact that -1 is not a local norm at 2 (Theorem 10.3).

Lemma 1. $\operatorname{Num}_q(E_q, o^*_{q,1}) = \operatorname{Num}_q(E_q, o^*_{q,-1}) = \frac{1}{2} \operatorname{Num}_q^+$.

Proof. We have

$$h_{E_q}(t,s) = h_{E_q}(-t,s)$$

because the map $\sigma \mapsto -\sigma$ reverses traces, does not change the determinant, and is measure preserving. Therefore

$$\int_{o^*_{q,1}} h_{E_q}(\operatorname{Tr} z, Nz)\, dz = \int_{o^*_{q,1}} h_{E_q}(-\operatorname{Tr} z, Nz)\, dz$$

$$= \int_{o^*_{q,-1}} h_{E_q}(\operatorname{Tr} z, Nz)\, dz.$$

This comes from $-o^*_{q,1} = o^*_{q,-1}$, because -1 is not a square in $Z(q)^*$, and hence represents the coset of non-squares in the residue class field of o_q, which is the same as $Z(q)$. Since $o^*_{q,1} \cup o^*_{q,-1} = o^*_q$, the lemma follows from the definition of Num_q^+.

Let us define

$$Y_\zeta = o^*_{2,\zeta} \cup o^*_{2,-\zeta}, \quad X_\zeta = E_{2,\zeta} \cup E_{2,-\zeta}.$$

The decomposition (2) and Lemma 1 give:

(3) $$\operatorname{Num}_{2q}(E_{2q,\zeta}, o^*_{2q,\zeta}) = \frac{1}{2} \operatorname{Num}_q^+ \operatorname{Num}_2(E_{2,\zeta}, Y_\zeta),$$

and (1) shows that

$$\mathrm{Num}_{2q}(\tilde{\mathcal{G}}_{2q}) = \sum_\zeta \mathrm{Num}_{2q}(E_{2q,\zeta}, \mathrm{o}^*_{2q,\zeta})$$

$$= \tfrac{1}{2} \mathrm{Num}_q^+ \sum_\zeta \mathrm{Num}_2(E_{2,\zeta}, Y_\zeta)$$

$$= \tfrac{1}{2} \mathrm{Num}_q^+ \sum_{\zeta = 1, \omega, \omega^2} \mathrm{Num}_2(X_\zeta, Y_\zeta).$$

We must therefore compute each term of this last sum. Observe that each such term is invariant under $\zeta \mapsto -\zeta$.

By Part II, Theorem 10.2 we know that $\mathrm{Num}_q^+ = 1 - 1/q$, because we are in the ramified case. The appropriate integrals will be computed to give:

Theorem 12.1. *Let the assumptions be those stated at the beginning of the section. Then*

$$\mathrm{Num}_{2q}(\tilde{\mathcal{G}}_{2q}) = \tfrac{11}{48} \tfrac{1}{2} (1 - 1/q).$$

Proof. We use Theorem 2.4. As in Theorem 11.1, let

$$\tau = \begin{pmatrix} -1 & 2 \\ 2 & 1 \end{pmatrix} \quad \text{and} \quad \gamma = \begin{pmatrix} -1 & -1 \\ 1 & 0 \end{pmatrix}.$$

Since we work at the prime 2, we omit the subscript 2, and write E_ζ instead of $E_{2,\zeta}$. By our assumption on τ, we get

$$E'(4) = (I + 2O(2)) \cup I$$

$$E_1(4) = E'(4) \cup E'(4)\tau.$$

The elements of $O(2)$ have been tabulated (cf. Theorem 2.4), and all have trace $\equiv 0 \pmod{2}$. The group E_1 is by definition the inverse image of $E_1(4)$ under reduction mod 4. We then get

$$E = X_1 \cup \gamma X_1 \cup \gamma^2 X_1.$$

We have to tabulate the trace and determinant properties (mod 4) for the three sets X_1, γX_1, $\gamma^2 X_1$. We note that

$$X_1(4) = \pm E_1(4) \cup \pm E_1(4)\tau.$$

The table follows, giving the trace and determinant mod 4.

		trace	det
$X_1(4) = $	$\pm I, \pm I + 2O(2)$	2	1
	$\pm r, \pm r + 2O(2)$	0	-1
$\gamma X_1(4) = $	$\gamma, \gamma + 2O(2)$	-1	1
	$-\gamma, -\gamma + 2O(2)$	1	1
	$\gamma r, \gamma r + 2O(2)$	1	-1
	$-\gamma r, -\gamma r + 2O(2)$	-1	-1
$\gamma^2 X_1(4) = $	$\gamma^2, \gamma^2 + 2O(2)$	-1	1
	$-\gamma^2, -\gamma^2 + 2O(2)$	1	1
	$\gamma^2 r, \gamma^2 r + 2O(2)$	-1	-1
	$-\gamma^2 r, -\gamma^2 r + 2O(2)$	1	-1

We observe that Y_1 has index 3 in o_2^*, and consists of the cubes, whence

$$Y_1 = 1 + 2o_2 .$$

The cosets are represented by $1, \omega, \omega^2$. We have

$$X_\zeta = \gamma X_1 \quad \text{or} \quad \gamma^2 X_1, \quad \text{if} \quad \zeta = \omega \quad \text{or} \quad \omega^2,$$

and it is not necessary for us to know which.

We now compute the desired terms for $\zeta = \omega$ or ω^2.

From the table, we see that the elements of γX_1 and $\gamma^2 X_1$ are non scalar mod 2. For each one of these sets, every combination of 1 and -1 occurs, the same number of times, namely 4 times. From Part II, Theorem 7.1 we find:

$$h_{\gamma X_1}(\text{Tr } z, Nz) = h_{\gamma^2 X_1}(\text{Tr } z, Nz) = \begin{cases} \frac{1}{4} & \text{if Tr } z \text{ is odd,} \\ & Nz \text{ is a unit} \\ 0 & \text{otherwise.} \end{cases}$$

All elements of Y_ω or Y_{ω^2} have odd trace, and their norm is a unit. Consequently for $\zeta = \omega$ or ω^2 we get

(4) $$\text{Num}_2(X_\zeta, Y_\zeta) = \tfrac{1}{4}\mu(Y_\zeta) = \tfrac{1}{12}\mu(o_2^*) = \tfrac{1}{16}.$$

Finally, we compute the term with $\zeta = 1$.

All elements of Y_1 have even trace, and the elements of o_2^* which are not in Y_1 have odd trace. All elements of X_1 have even trace. Hence

$$\text{Num}_2(X_1, Y_1) = \text{Num}_2(X_1, o_2^*) = \text{Num}_2(X_1, o_2^*).$$

This leaves us with four integrals to compute. By Part II, §8, Lemma 6,

(5) $$\int_{o_2} h_{I+4M}(\text{Tr } z, Nz)\,dz = 5/384.$$

This integral will be counted 2 times.

Next, let $\sigma \in O_2$. We use constantly Part II, §8, Lemmas 4, 5, 6 without further reference. We obtain

(6) $$\int_{o_2} h_{I+2\sigma+4M}(\text{Tr } z, Nz)\,dz = \tfrac{1}{8} \int_{1+2o_2} h_{\sigma+2M}$$
$$= \tfrac{r^2}{8}\mu(1+2o_2)$$
$$= \tfrac{1}{128}.$$

This integral will be counted 6 times.

Next we write $\tau = I + 2\gamma \pmod 4$. We use Part II, Theorem 7.1. We get:

(7) $$\int_{o_2} h_{\tau+4M}(\text{Tr } z, Nz)\,dz = \tfrac{1}{8}\int_{o_2} h_{\gamma+2M}(\text{Tr } z, Nz)\,dz$$
$$= \tfrac{1}{8}\tfrac{1}{4}[\mu(\omega+2o_2) + \mu(\omega^2+2o_2)]$$
$$= 1/64.$$

This integral will be counted 2 times.

Next and last, we get

(8) $$\int_{\mathfrak{o}_2} h_{\tau+2\sigma+4M}(\text{Tr } z, Nz)\,dz = \frac{1}{8} \int_{\mathfrak{o}_2} h_{\gamma+\sigma+2M}(\text{Tr } z, Nz)\,dz$$
$$= 0.$$

because $\gamma + \sigma$ has odd trace and even determinant, which is incompatible with any element of $\mathfrak{o}_2$.

Taking the sum of the last four integrals appropriately weighted yields

(9) $$\text{Num}_2(X_1, Y_1) = \frac{2 \cdot 5}{384} + \frac{6}{128} + \frac{2}{64} = \frac{5}{3} \cdot \frac{1}{16}.$$

Hence by (4) and (9) we obtain

$$\sum_{\zeta=1,\omega,\omega^2} \text{Num}_2(X_\zeta, Y_\zeta) = \frac{2}{16} + \frac{5}{3} \cdot \frac{1}{16} = \frac{11}{48}.$$

This concludes the proof.

PART IV

NUMERICAL RESULTS

PART IV

NUMERICAL RESULTS

SUPERSINGULAR AND FIXED TRACE DISTRIBUTION

1. General discussion of results — 235
2. Tables
   - Table I : Fixed trace distributions — 239
   - Table II : Supersingular primes — 240
   - Table III: Primes with $t_p = 1$ — 241
   - Table IV: Traces of Frobenius — 242

IMAGINARY QUADRATIC DISTRIBUTION

3. General discussion of results — 249
4. Tables
   - Table V : Imaginary quadratic distributions — 253
   - Table VI : Primes associated with fields of small discriminant, for curves A and B — 258
   - Table VII: Distribution of primes associated with small discriminants — 260

EXTENDED RESULTS FOR $X_0(11)$

5. Discussion and description of tables — 265
   - Table VIII: Supersingular primes — 267
   - Table IX : Imaginary quadratic distribution — 268
   - Table X : Distribution of primes for fields with small discriminants — 269

Remarks on the Computations — 271

Bibliography — 273

SUPERSINGULAR AND FIXED TRACE DISTRIBUTION

§1. General discussion of results

In this final part we present the results of numerical calculations for five curves. Four of them, which are arbitrarily labeled A, B, C, D appear in Serre [S 2] as examples 5.5.6, 5.5.7, 5.5.8 and 5.9.2. They have the property that the group G is of index 2 in $\prod GL_2(Z_\ell)$. See §5 and §7 of Part I. The fifth is the modular curve $X_0(11)$ whose Galois group is determined in §8, and which also appears in [S 2] as example 5.5.2.

The following table gives the equation, conductor N, discriminant Δ, and j-invariant for each curve.

		N	Δ	j
A	$y^2 + y = x^3 - x$	37	37	$2^{12}3^3/37$
B	$y^2 + y = x^3 + x^2$	43	-43	$-2^{12}/43$
C	$y^2 + xy + y = x^3 - x^2$	53	53	$-3^35^3/53$
D	$y^2 = x^3 + 6x - 2$	2^63^3	-2^63^5	2^93
$X_0(11)$	$y^2 + y = x^3 - x^2 - 10x - 20$	11	-11^5	$-2^{12}31^3/11^5$

We used the machine to compute the values of the Frobenius traces t_p for each of these curves for the first 5,000 primes, from $p_1 = 2$ to $p_{5,000} = 48,611$. More extensive calculations for the curve $X_0(11)$ are reported in §5. We also calculated the constant $C(t, A)$ for $-7 \leq t \leq 7$ given by the formulas developed in §5 and §8 of Part I.

In table I, the first column for each curve gives the number of primes among the first 5,000 with the given trace. The second column gives the value predicted by our conjecture, i.e.

$$E_t = E_t(48,611) = C(t, A)\pi_{1/2}(48,611) = C(t, A) 26.434 ,$$

where the value of $\pi_{1/2}(48,611)$ is obtained by direct calculation. The number in the third column indicates roughly how well the data fit the conjecture for each particular case. Its meaning is explained further below. In statistical terms, the overall fit is quite good. The standard χ^2-test involves calculating the sum of

$$\frac{(\text{Predicted value} - \text{Actual value})^2}{\text{Predicted value}}$$

over all the cases. For the data in table I, we get the following results (omitting the traces $\equiv 1 \mod 5$ for $X_0(11)$ where the predicted value is 0).

Curve	χ^2-value	Probability
A	8.4	.91
B	14.8	.46
C	16.8	.32
D	10.8	.75
$X_0(11)$	5.6	.93

The column headed "probability" gives the chance that a random sample using the theoretical frequencies would give a χ^2-value greater than that observed, i.e. would give no better fit. (The values are from standard statistical tables, using 15 degrees of freedom for A, B, C, and D where there are 15 possibilities, and 12 degrees of freedom for $X_0(11)$.) By this criterion there is very good fit for A and $X_0(11)$, and reasonably good fit in all cases. The single prime that has $t_p = 1$ for $X_0(11)$ is the prime 5, which (unlike all other primes) is congruent to 0 modulo 5.

The third column for each curve in table I is calculated as follows. For a truly random model, the probability of getting a count of k when the theoretical frequency is m is well approximated by the formula for the Poisson distribution

$$p_k = e^{-m} m^k / k! \ .$$

The value given in the table is $\sum p_i$, with the sum taken over all i such that $p_i \leq p_k$. Thus the value in the table is the chance that a random experiment based on the given theoretical frequency would not give a more likely result. Note that the value is 1 (printed as .99) if m comes anywhere between k and $k+1$, since then p_k is the maximum of the p_i. No significance should be given to the precise value printed, but it does give a reasonable indication of how well the data agree with the prediction.

There are some rather low values, at traces −5 and 1 for B, and traces −6 and 4 for C, where the probability is about 1 in 30. It should be noted, however, that having as many as four events, each with probability about 1/30, occurring in a total of 72 "experiments," is not unlikely. Somewhat more disturbing is that in 8 out of 9 cases with trace ±4, the predicted value is noticeably less than the actual value. We don't see any easy way of pushing computations so much farther as would make it clear if this phenomenon persists or disappears.

In table II we list for each curve the supersingular primes $\leq 48,611$ (that is, those among the first 5,000 primes), and in table III, the primes with $t_p = 1$, which are of special interest in the light of Mazur's work [Ma]. Table IV shows the values of t_p for the first 200 primes.

Table I

t	A N_t	E_t	Pr	B N_t	E_t	Pr	C N_t	E_t	Pr
-7	7	10.3	.43	10	10.4	.99	14	10.3	.27
-6	22	24.9	.69	25	24.8	.92	14	24.8	.03
-5	13	10.3	.35	4	10.4	.04	13	10.3	.35
-4	26	20.7	.23	26	20.7	.23	25	20.7	.32
-3	14	12.4	.57	10	12.4	.67	12	12.4	.99
-2	18	20.7	.66	18	20.7	.66	21	20.7	.91
-1	13	10.3	.53	13	10.4	.35	14	10.3	.27
0	21	27.3	.25	23	28.0	.39	28	27.4	.85
1	10	10.3	.99	18	10.4	.03	6	10.3	.21
2	21	20.7	.91	20	20.7	.99	20	20.7	.99
3	11	12.4	.89	12	12.4	.99	11	12.4	.89
4	25	20.7	.32	24	20.7	.44	11	20.7	.03
5	10	10.3	.99	11	10.4	.76	7	10.3	.43
6	19	24.9	.27	27	24.8	.62	24	24.8	.99
7	9	10.3	.88	13	10.4	.35	9	10.3	.88

t	D N_t	E_t	Pr	$X_0(11)$ N_t	E_t	Pr
-7	9	12.4	.40	13	13.2	.99
-6	28	29.0	.99	35	31.2	.47
-5	10	13.1	.49	10	12.5	.57
-4	21	18.6	.56	0	0.0	--
-3	7	8.3	.86	21	15.8	.21
-2	19	18.6	.91	23	26.0	.62
-1	9	12.4	.40	12	13.2	.89
0	33	32.3	.86	33	34.7	.87
1	12	12.4	.99	1	0.0	--
2	24	18.6	.20	24	26.0	.77
3	5	8.3	.38	14	15.8	.80
4	25	18.6	.16	33	26.0	.17
5	10	13.1	.49	12	12.5	.99
6	24	29.0	.40	0	0.0	--
7	16	12.4	.32	12	13.2	.89

Fixed trace distributions

Table II

A	B	C	D	$X_0(11)$
17	7	5	101	19
19	37	11	251	29
257	1109	239	1637	199
311	1361	751	2383	569
577	1531	1201	2411	809
2161	2069	1459	2441	1289
2243	2281	1663	3631	1439
3511	2543	2089	4241	2539
7951	3011	2111	4889	3319
10399	5351	2153	8081	3559
11261	6569	2411	10979	3919
11579	7211	3001	11399	5519
15331	9923	4283	11471	9419
21031	12289	5443	12503	9539
21991	16339	5867	13577	9929
23369	19819	6113	15053	11279
23567	21191	8999	15131	11549
30319	29741	9127	15887	13229
31481	31531	9227	16619	14489
38167	31547	10079	17417	17239
40127	32411	10667	19913	18149
	38711	12071	22271	18959
	41077	15269	23057	19319
		29221	23561	22279
		32467	23981	24359
		33619	25163	27529
		45613	27799	28789
		46817	28439	32999
			30839	33029
			33767	36559
			35591	42899
			41399	45259
			46307	46219

Supersingular primes ≤ 48,611

Table III

A	B	C	D	$X_o(11)$
53	103	71	7	5
127	127	97	97	
443	541	1063	1447	
599	1429	3061	1663	
3989	1657	4993	6277	
23269	2087	5903	6397	
29131	3733		7039	
30529	8641		8821	
44179	8753		13147	
47237	10903		13417	
	18313		15919	
	21467		19333	
	34739			
	34897			
	41453			
	44579			
	46261			
	47857			

Primes ≤ 48,611 with $t_p = 1$

Table IV Traces of Frobenius

p	A	B	C	D	$X_o(11)$
2	-2	-2	-1	*	-2
3	-3	-2	-3	*	-1
5	-2	-4	0	2	1
7	-1	0	-4	1	-2
11	-5	3	0	2	*
13	-2	-5	-3	-1	4
17	0	-3	-3	-6	-2
19	0	-2	-5	5	0
23	2	-1	7	6	-1
29	6	-6	-7	8	0
31	-4	-1	4	8	7
37	*	0	5	5	3
41	-9	5	6	-8	-8
43	2	*	-2	4	-6
47	-9	4	-2	-10	8
53	1	-5	*	4	-6
59	8	-12	-2	-14	5
61	-8	2	-8	-3	12
67	8	-3	-12	13	-7
71	9	2	1	-4	-3
73	-1	2	-4	9	4
79	4	-8	-1	11	-10
83	-15	15	-1	12	-6
89	4	-4	-14	2	15
97	4	7	1	1	-7
101	3	-9	-2	0	2
103	18	1	-1	-3	-16
107	-12	-12	6	-2	18
109	-16	7	16	10	10
113	-18	-20	15	10	9
127	1	1	13	20	8
131	-12	8	-2	-16	-18
137	-6	6	12	6	-7
139	4	19	-20	-5	10
149	-5	12	-5	-12	-10
151	16	-20	-3	-9	2
157	23	-10	-4	-2	-7
163	-18	14	-6	11	4
167	-12	-9	21	14	-12
173	9	6	10	-24	-6
179	18	20	11	12	-15
181	5	10	-2	19	7
191	-4	-16	-21	-6	17
193	-26	3	-16	-21	4
197	3	2	-18	2	-2
199	2	14	4	-19	0
211	-13	2	-2	-1	12
223	-17	-28	-14	16	19
227	-16	-4	6	-12	18
229	7	-15	21	-26	15

* indicates bad reduction

Table IV Traces of Frobenius (continued)

p	A	B	C	D	$X_o(11)$
233	6	6	-8	24	24
239	-6	16	0	12	-30
241	14	-12	-11	13	-8
251	-2	-23	20	0	-23
257	0	-24	-28	4	-2
263	19	-18	-28	-12	14
269	-6	-25	9	6	10
271	-31	23	-14	-21	-28
277	12	-32	-8	2	-2
281	12	19	17	-24	-18
283	4	21	-9	-20	4
293	-2	-26	26	-18	24
307	-17	-7	-16	12	8
311	0	15	16	-30	12
313	22	22	-6	-17	-1
317	22	9	3	12	13
331	-2	-26	28	-29	7
337	-25	-3	10	-31	-22
347	-10	28	-30	4	28
349	6	14	24	5	30
353	8	-31	-18	-4	-21
359	-15	19	9	-14	-20
367	8	-32	22	-15	-17
373	-19	32	-10	-11	-26
379	15	11	11	-19	-5
383	20	32	16	-16	-1
389	4	6	12	18	-15
397	-5	-6	18	14	-2
401	18	5	-32	-28	2
409	20	-24	19	13	-30
419	7	-28	-12	26	20
421	-24	-10	14	19	22
431	-30	-21	-36	30	-18
433	9	-12	-7	-2	-11
439	28	17	-34	-8	40
443	1	-4	-33	-36	-11
449	36	30	-11	-34	35
457	18	-18	38	2	-12
461	30	30	-9	30	12
463	-22	4	-23	-19	-11
467	-2	6	-28	2	-27
479	14	21	3	12	20
487	-24	36	20	-11	23
491	-28	-6	27	14	-8
499	12	-8	-23	24	20
503	16	6	16	6	-26
509	-31	-15	2	-6	15
521	-33	14	-45	-42	-3
523	-22	12	-42	-15	-16
541	20	1	22	-37	-8

Table IV Traces of Frobenius (continued)

p	A	B	C	D	$X_o(11)$
547	8	-29	-38	15	8
557	-18	-3	14	14	-2
563	-30	37	-1	24	4
569	-24	7	24	10	0
571	7	-14	-4	-19	-28
577	0	-20	10	3	33
587	-32	2	28	6	28
593	-5	-16	25	36	44
599	1	-1	-30	8	40
601	-22	-4	22	-26	2
607	-32	-4	-8	41	-22
613	15	-18	-16	9	-16
617	17	-21	-42	-42	18
619	-1	36	16	9	-25
631	-28	-6	7	-1	7
641	-1	12	46	-4	-33
643	14	-36	-8	12	29
647	-8	-12	38	24	-7
653	-24	-14	-38	-44	-41
659	-15	-19	12	-36	10
661	-28	31	-1	-23	37
673	27	14	-10	23	14
677	-11	34	40	-14	-42
683	18	-9	2	-24	-16
691	-20	-40	41	-4	17
701	-12	-2	12	30	2
709	40	-1	6	23	-25
719	39	8	24	-52	15
727	16	16	-30	48	3
733	7	32	14	34	-36
739	-9	-10	15	20	50
743	21	-24	-26	-42	4
751	25	-6	0	-25	-23
757	-50	28	-43	-33	-22
761	-35	20	-28	42	12
769	26	-42	-2	-17	20
773	-9	-4	40	-36	-6
787	-5	4	-4	-3	-32
797	52	-42	16	-24	53
809	2	26	-16	20	0
811	47	-14	42	20	-38
821	-47	49	-24	2	22
823	-16	-1	26	43	39
827	22	-36	47	2	-52
829	-4	44	-30	-41	25
839	44	-40	-30	-4	-5
853	26	-29	-56	-15	14
857	-48	-10	42	-28	8
859	-20	-32	-58	19	-15
863	-24	-6	-30	38	24

Table IV Traces of Frobenius (continued)

p	A	B	C	D	$X_o(11)$
877	50	41	-2	17	-12
881	-14	37	16	6	-43
883	48	31	-36	31	4
887	25	-22	-7	12	-22
907	52	47	46	17	-12
911	26	-22	-32	-48	12
919	-58	-49	15	48	10
929	18	-6	38	-12	-30
937	37	32	-55	-25	8
941	-10	33	10	22	42
947	12	-33	-12	6	-27
953	61	22	-42	-6	34
967	-14	37	52	27	-32
971	-8	-13	42	-34	47
977	28	34	-20	-44	-27
983	9	-24	46	42	39
991	-18	2	34	-7	-8
997	-42	4	-13	54	38
1009	-47	-18	16	-11	-10
1013	36	-22	39	-26	39
1019	46	-30	15	60	-10
1021	-62	-14	-52	54	22
1031	-4	46	-22	22	32
1033	3	13	-54	-59	-16
1039	-59	34	44	-23	5
1049	-4	6	3	20	-55
1051	-16	-60	-30	4	2
1061	-62	-40	-31	54	-13
1063	7	12	1	31	44
1069	-30	32	-23	-46	-20
1087	12	20	-25	59	8
1091	30	40	-13	20	-58
1093	-36	-46	-14	-22	-51
1097	36	12	-1	-2	-42
1103	-8	54	-26	24	-51
1109	-35	0	-11	28	-30
1117	33	4	-53	-45	48
1123	-22	60	-52	-41	24
1129	50	-26	-5	41	50
1151	-25	-30	10	-18	2
1153	18	21	6	37	-31
1163	-36	32	-12	44	34
1171	-22	-52	11	-53	-3
1181	57	-40	55	22	-18
1187	-33	-48	3	-26	-12
1193	-11	32	-54	-18	-21
1201	44	-55	0	-5	2
1213	-12	-59	31	47	-41
1217	-41	-30	48	-14	-42
1223	30	-6	-12	-4	14

IMAGINARY QUADRATIC DISTRIBUTION

§3. General discussion of results

For each of the curves already described and each of the first 5,000 primes, we used the machine to find the square-free part of $4p - t_p^2$, and thus determine the quadratic field k, (excluding supersingular primes and those with bad reduction).

Table V gives data for each curve on the number of primes found for the 31 fields of discriminant D with $|D| < 100$. Two sets of data are given for each curve. The first refers to primes among the first 5,000, and the second to primes p such that
$$p_{72} = 359 < p \leqq p_{5,000} = 48,611 .$$

Our conjecture deals with asymptotic behavior, and our heuristic arguments involve the assumption that p is large compared to $|D|$. It is clearly very unlikely for large values of D to occur with small p, but small values of p contribute heavily to the factor
$$\pi_{1/2}(x) = \sum_{p \leq x} \frac{1}{2\sqrt{p}} .$$

We therefore thought it appropriate to look at the result of eliminating some of the small primes from consideration. Choosing to discard exactly 72 primes was arbitrary, but convenient because we had already subdivided the data in that way (see the description of Table VII below).

The table has a page for each curve. The first column gives the field discriminant, the next three pertain to the first 5,000 primes, and the last three to the primes with the first 72 primes omitted. In each group of 3 columns, the first gives the actual number of primes found, the second gives the expected number E_D calculated from the formulas of Parts II and III, and the third gives the probability that a random experiment would not give a more likely result. (See the discussion of these probabilities given in §1.)

As in the fixed trace case, there is generally good agreement between actual and predicted counts. There is a very conspicuous discrepancy at $D = -11$ for the curve $X_0(11)$, when all primes are considered, but the discrepancy becomes much less significant when the low primes are discarded. There are several other cases where the probability falls as low as .01, but it is statistically reasonable for a few such low probabilities to come up among 154 cases. (See §5 for further discussion of the curve $X_0(11)$.)

A χ^2 computation using the data on only the first 15 discriminants (to avoid excessively small frequencies) gives the following values. (See the discussion in §1 for an explanation. There are 15 degrees of freedom, except for curve D, which has 14 because the predicted value for the field $Q(\sqrt{-3})$ was not calculated.)

	5,000 primes		4,928 primes	
Curve	χ^2	Probability	χ^2	Probability
A	17.8	.27	16.9	.32
B	7.8	.93	8.0	.92
C	15.6	.41	15.3	.44
D	7.1	.93	7.5	.91
$X_0(11)$	26.7	.03	14.6	.47

By this criterion, the fit is reasonably good except for $X_0(11)$ when all 5,000 primes are considered. This high value for χ^2 can be attributed to the bad fit at $D = -11$, and when the small primes are discarded, the result is quite acceptable.

For most entries in the table, we have $K \cap k_{ab} = Q_{ab}$, and the predicted values are found from the formulas of Part II. It was shown in Part III that the same formulas apply for $k = Q(i)$, and if $q \equiv 1 \pmod 3$, for $k = Q(\sqrt{-3})$, even though

$$K \cap k_{ab} \neq Q_{ab}$$

in these cases. For the curve C and $X_0(11)$, with $k = Q(\sqrt{-3})$, we have

$$q \equiv -1 \pmod 3,$$

so the formulas of Part III apply. The value obtained, however, differs by less than one per cent from the value one gets simply applying the formulas of Part II.

The situation is very different for the curve B, with $D = -43$, and the curve $X_0(11)$, with $D = -11$, where $k = Q(\sqrt{\Delta})$. Taking the intersection $K \cap k_{ab}$ fully into account increases the prediction by a factor of almost 2. For $D = -43$

there is a sharp difference between the actual values for the curve B and the other Serre curves, perfectly in line with the prediction. For $D = -11$ and $X_0(11)$, the situation is further complicated by the factor at 5, see Theorem 11.3 of Part II.

Except for $k = Q(\sqrt{-43})$, one notices that the predicted values for A, B, C are identical to the first decimal place. This is not surprising, because the tables of §10 show that the constants differ by a factor of the form $1 + O(1/q)$, and the values of q for the three curves are 37, 43, 53. The value of q for curve D is 3, so slight differences appear.

In Table VI we give a complete list of all the primes corresponding to the indicated discriminants for the curve A and B. Visually, the table looks like a parabola with kinks, as it should. The other cases we have computed look roughly the same, so we did not feel it was worthwhile to include them.

In Table VII we present a more detailed breakdown of how the primes belonging to different fields are distributed. If one divides the primes up to $p_{5,000} = 48{,}611$ into 12 segments

$$X_i = \{p, p_{i-1} < p \leq p_i\}$$

so as to make the sums

$$\sum_{p \in X_i} \frac{1}{2\sqrt{p}}$$

as nearly equal as possible, the division points come out as follows.

Segment	i	p_i
1	16	53
2	72	359
3	187	1117
4	371	2539
5	632	4673
6	976	7691
7	1407	11719
8	1930	16691
9	2548	22807
10	3263	30169
11	4080	38713
12	5000	48611

According to our conjecture, one would have the primes associated with a given field equally distributed over the 12 segments. Of course, since the number predicted for an individual segment is small, considerable statistical fluctuation is to be expected.

In the table, the first column gives the field discriminant, and the second gives the total count for each field. The last 12 columns give the counts $N_D(X_i)$ for the individual segments. The bottom line gives the total counts for each segment. The total count for the first segment is noticeably low, which is not surprising since it contains only 16 primes altogether. Otherwise, the distribution across segments appears reasonably uniform.

We discarded the first two segments to get the reduced data used for the second group of columns in Table V. Since the sums

$$\sum_{p \in X_i} \frac{1}{2\sqrt{p}}$$

are almost equal for $i = 1, 2, \cdots, 12$, the theoretical frequencies for the reduced set of primes are very nearly $10/12$ of the corresponding frequencies for the full set.

Table V

D	N_D	E_D	Pr	N_D	E_D	Pr
-3	40	34.9	.40	30	29.1	.85
-4	24	28.5	.45	21	23.7	.68
-7	18	17.0	.81	18	14.2	.29
-8	21	18.0	.48	18	15.0	.44
-11	17	15.8	.71	16	13.2	.41
-15	13	10.9	.54	12	9.1	.32
-19	13	13.6	.99	12	11.3	.77
-20	19	10.9	.02	15	9.1	.06
-23	6	8.1	.60	6	6.7	.99
-24	5	10.7	.09	4	8.9	.13
-31	4	7.5	.27	3	6.3	.31
-35	5	8.9	.24	5	7.4	.46
-39	8	6.5	.55	8	5.4	.27
-40	7	8.9	.74	6	7.4	.85
-43	14	9.9	.20	13	8.2	.11
-47	2	5.5	.19	2	4.6	.34
-51	10	8.0	.48	9	6.7	.33
-52	10	8.0	.48	10	6.7	.18
-55	2	5.8	.14	2	4.8	.26
-56	8	6.3	.43	7	5.3	.38
-59	7	6.6	.84	7	5.5	.52
-67	10	8.1	.48	9	6.7	.33
-68	7	5.9	.54	7	4.9	.36
-71	2	4.3	.46	1	3.6	.28
-79	3	4.7	.64	3	4.0	.99
-83	3	5.8	.30	2	4.9	.26
-84	4	5.7	.67	4	4.8	.99
-87	1	4.4	.15	0	3.7	.06
-88	5	6.4	.84	2	5.4	.19
-91	7	6.4	.69	7	5.3	.39
-95	4	3.8	.80	3	3.2	.99

Curve A

Table V

D	N_D	E_D	Pr	N_D	E_D	Pr
-3	34	34.9	.99	29	29.1	.99
-4	26	28.5	.71	23	23.7	.99
-7	18	17.0	.81	16	14.2	.59
-8	14	18.0	.41	9	15.0	.15
-11	12	15.8	.45	12	13.2	.89
-15	8	10.9	.45	7	9.1	.62
-19	9	13.6	.28	8	11.3	.45
-20	10	10.9	.99	9	9.1	.99
-23	5	8.1	.38	4	6.7	.44
-24	12	10.7	.64	11	8.9	.40
-31	4	7.5	.27	3	6.3	.31
-35	8	8.9	.99	7	7.4	.99
-39	7	6.5	.70	6	5.4	.67
-40	9	8.9	.87	9	7.4	.58
-43	23	25.1	.76	20	20.9	.99
-47	1	5.5	.05	1	4.6	.10
-51	13	8.0	.11	12	6.7	.05
-52	8	8.0	.99	7	6.7	.85
-55	6	5.8	.83	6	4.8	.49
-56	9	6.3	.31	8	5.3	.27
-59	6	6.6	.99	5	5.5	.99
-67	10	8.1	.48	10	6.7	.24
-68	1	5.9	.04	1	4.9	.11
-71	4	4.3	.99	3	3.6	.99
-79	7	4.7	.25	6	4.0	.30
-83	4	5.8	.68	4	4.9	.99
-84	1	5.7	.05	0	4.8	.02
-87	3	4.4	.81	3	3.7	.99
-88	2	6.4	.11	2	5.4	.19
-91	6	6.4	.99	5	5.3	.99
-95	4	3.8	.80	4	3.2	.57

Curve B

Table V

D	N_D	E_D	Pr	N_D	E_D	Pr
-3	37	34.9	.67	33	29.1	.46
-4	16	28.5	.01	15	23.7	.08
-7	13	17.0	.40	12	14.2	.69
-8	13	18.0	.29	11	15.0	.37
-11	15	15.8	.99	15	13.2	.58
-15	12	10.9	.65	11	9.1	.50
-19	13	13.6	.99	11	11.3	.99
-20	8	10.9	.45	5	9.1	.24
-23	5	8.1	.38	5	6.7	.70
-24	14	10.7	.28	14	8.9	.09
-31	7	7.5	.99	5	6.3	.84
-35	6	8.9	.40	5	7.4	.46
-39	8	6.5	.55	7	5.4	.51
-40	4	8.9	.13	3	7.4	.14
-43	12	9.9	.43	9	8.2	.73
-47	9	5.5	.13	9	4.6	.05
-51	8	8.0	.99	7	6.7	.85
-52	3	8.0	.08	3	6.7	.24
-55	3	5.8	.40	3	4.8	.64
-56	7	6.3	.69	7	5.3	.38
-59	10	6.6	.17	9	5.5	.13
-67	10	8.1	.48	8	6.7	.56
-68	8	5.9	.40	6	4.9	.65
-71	2	4.3	.46	2	3.6	.60
-79	5	4.7	.82	5	4.0	.61
-83	4	5.8	.68	4	4.9	.99
-84	6	5.7	.83	6	4.8	.49
-87	5	4.4	.64	5	3.7	.43
-88	7	6.4	.69	7	5.4	.39
-91	10	6.4	.16	10	5.3	.05
-95	1	3.8	.20	1	3.2	.39

Curve C

Table V

D	N_D	E_D	Pr	N_D	E_D	Pr
-3	46	--	--	43	--	--
-4	27	26.3	.84	23	21.9	.75
-7	14	17.0	.54	13	14.2	.89
-8	20	18.0	.64	18	15.0	.44
-11	14	15.8	.80	13	13.2	.99
-15	11	10.9	.88	9	9.1	.99
-19	13	14.2	.89	13	11.9	.66
-20	11	10.9	.88	11	9.1	.50
-23	5	8.1	.38	5	6.7	.70
-24	4	5.3	.83	4	4.4	.99
-31	4	7.5	.27	4	6.3	.54
-35	9	8.9	.87	9	7.4	.58
-39	4	6.5	.43	2	5.4	.19
-40	9	8.2	.73	7	6.8	.85
-43	6	10.4	.21	4	8.6	.13
-47	7	5.5	.52	6	4.6	.48
-51	4	10.5	.04	2	8.8	.02
-52	4	7.4	.27	3	6.2	.31
-55	1	5.8	.04	1	4.8	.10
-56	1	6.3	.03	0	5.3	.01
-59	10	6.6	.17	8	5.5	.28
-67	7	8.5	.86	5	7.1	.57
-68	7	5.9	.54	7	4.9	.36
-71	6	4.3	.34	6	3.6	.18
-79	2	4.7	.35	2	4.0	.45
-83	4	5.8	.68	4	4.9	.99
-84	3	2.9	.77	2	2.4	.99
-87	7	4.4	.22	7	3.7	.11
-88	9	5.9	.21	6	4.9	.65
-91	5	6.7	.70	5	5.6	.99
-95	4	3.8	.80	4	3.2	.57

Curve D

Table V

D	N_D	E_D	Pr	N_D	E_D	Pr
-3	29	35.5	.31	26	29.6	.58
-4	67	84.9	.05	58	70.7	.14
-7	15	17.2	.72	14	14.3	.99
-8	25	18.3	.13	23	15.2	.05
-11	88	120.3	.0026	84	100.2	.11
-15	13	13.1	.99	13	10.9	.54
-19	32	40.8	.18	29	33.9	.44
-20	11	13.0	.68	11	10.9	.88
-23	5	8.2	.38	5	6.8	.70
-24	26	32.0	.33	22	26.6	.44
-31	19	22.5	.53	15	18.8	.49
-35	8	10.7	.54	8	8.9	.99
-39	15	19.4	.36	14	16.2	.71
-40	5	10.7	.09	5	8.9	.24
-43	8	10.0	.64	8	8.3	.99
-47	3	5.6	.39	3	4.6	.64
-51	29	24.1	.31	28	20.1	.09
-52	6	8.1	.60	6	6.8	.99
-55	3	6.9	.18	3	5.8	.40
-56	15	18.9	.49	14	15.7	.80
-59	16	19.8	.50	15	16.5	.81
-67	8	8.2	.99	8	6.9	.57
-68	3	6.0	.30	3	5.0	.50
-71	11	13.0	.68	11	10.8	.88
-79	10	14.2	.35	10	11.9	.77
-83	4	5.9	.68	4	4.9	.99
-84	17	17.2	.99	15	14.3	.79
-87	2	4.5	.34	2	3.7	.60
-88	0	3.3	.09	0	2.7	.12
-91	21	19.2	.65	20	16.0	.32
-95	5	4.6	.81	5	3.8	.44

Curve $X_o(11)$

	-3	-4	-7	-8	-11	-15	-19	-20	-23	-24	-31	-35	-39	-40	-43
	3														
	7	2													
		5													
	13	317	487	113	157	139	11	29	2393	193	283	6521	367	251	59
	31	541	877	137	619	1021	19	61	2927	769	5701	7151	607	1019	509
	79	797	1373	179	1433	1549	853	109	8821	5281	17977	12301	2131	3251	821
	127	1061	2251	563	2689	2659	1109	281	9929	11113	37309	18691	2311	3449	3911
	211	2437	2543	1889	2777	3019	2063	401	13417	44119		33679	3121	17041	5431
	229	3089	3637	2897	2957	3391	4547	449	19553				6301	39769	6043
	241	3169	3691	3539	3617	12049	5573	2801					27901	44249	8807
	313	5581	4657	4139	4801	17491	5711	3221					28687		8839
	439	8893	5303	4409	11719	21961	5869	4441							9001
	733	11353	6469	4603	13183	26839	13687	5009							18289
	1039	14629	6911	6011	19429	29311	32491	5881							30931
	1231	15749	12517	6763	22639	32059	40189	8069							31147
	1753	17497	12671	8089	27241	35491	43963	12101							35081
	2671	19889	16349	13963	29297		45491	14549							39209
	3733	28657	19763	17449	30187			18121							
	4621	30113	20143	17939	45137			20921							
	4813	32233	22691	20113	47777			22409							
	5119	36493	33721	25321				26881							
	7549	36697		34273				35281							
	8161	39857		41771											
	8707	43321		42569											
	9403	47521													
	14479														
	14947														
	15937														
	16033														
	16231														
	21157														
	21517														
	31153														
	32713														
	33769														
	34303														
	38923														
	39667														
	41221														
	41941														
	44959														

Table VI A

Primes ≤ 48,611 associated with discriminants ≥ -43

for the curve A : $y^2 + y = x^3 - x$

-3	-4	-7	-8	-11	-15	-19	-20	-23	-24	-31	-35	-39	-40	-43
13	2	7	3											
127	5	79	19	421	61	251	29	59	199	293	11	181	809	47
223	197	239	73	1367	1129	6271	641	1871	3391	3187	709	673	4001	109
313	593	701	137	1423	1549	8689	1669	11069	6361	22483	2689	1249	4409	359
349	1489		331	1973	1831	14251	2729	11801	13879	45589	6091	33589	6211	977
1021	2833	2671	1409	10607	8329	15739	3001	16273	16231		10301	36919	9929	1559
1657	3049	2731	1723	13619	16921	21221	3821		31687		10691	39199	24281	1741
1801	3373	5261	2459	16831	34549	30841	8369		32719		19541	43627	24859	2003
2089	4933	5791	5153	18461	35521	34261	20089		33601		44809		31771	3881
2851	5981	6619	6113	19891		39251	28649		34033				33049	4357
3067	6397	8831	6833	24533			31741		37567					4813
3343	7349	12511	18059	26987					38791					6217
3919	8093	16349	29483	43591					48337					9431
6043	8761	17749	45827											11057
6121	10061	19867												17909
9067	12553	20147												18587
9199	13933	23143												18869
9391	15889	29327												20743
14389	19037	30319												30097
14983	23293	38851												39079
18013	24109													39929
18919	25057													41849
20479	28933													44893
21169	43577													45077
21751	43777													
24889	47057													
29803														
32353														
33409														
39409														
39631														
41269														
42649														
47809														

Table VI B

Primes ≤ 48,611 associated with discriminants ≥ −43

for the curve $B : y^2 + y = x^3 + x^2$.

Note the large frequency for $D = -43$.

Table VII

D	N_D											
					Curve A							
-3	40	4	6	3	2	3	3	3	5	2	0	4	5
-4	24	2	1	3	1	2	1	2	2	2	2	3	3
-7	18	0	0	2	2	4	3	0	3	3	0	1	0
-8	21	0	3	1	1	5	2	1	1	3	1	1	2
-11	17	0	1	1	1	4	1	1	1	2	2	1	2
-15	13	0	1	1	1	3	0	0	1	2	2	2	0
-19	13	1	0	2	1	1	3	0	1	0	0	1	3
-20	19	1	3	2	0	3	2	1	2	3	1	1	0
-23	6	0	0	0	1	1	0	2	1	1	0	0	0
-24	5	0	1	1	0	0	1	1	0	0	0	0	1
-31	4	0	1	0	0	0	1	0	0	1	0	1	0
-35	5	0	0	0	0	0	2	0	1	1	0	1	0
-39	8	0	0	2	2	1	1	0	0	0	2	0	0
-40	7	0	1	1	0	2	0	0	0	1	0	0	2
-43	14	0	1	2	0	1	2	3	0	1	0	3	1
Totals		8	19	21	12	30	22	14	18	22	10	19	19
					Curve B							
-3	34	1	4	1	3	4	2	3	2	5	2	2	5
-4	26	2	1	1	1	3	4	3	3	1	4	0	3
-7	18	0	2	1	0	2	3	1	2	3	2	1	1
-8	14	2	3	0	3	0	3	0	0	1	1	0	1
-11	12	0	0	1	3	0	0	1	1	3	2	0	1
-15	8	0	1	0	3	0	0	1	0	1	0	2	0
-19	9	0	1	0	0	0	1	1	2	1	0	2	1
-20	10	1	0	1	1	3	0	1	0	1	1	1	0
-23	5	0	1	0	1	0	0	1	2	0	0	0	0
-24	12	0	1	0	0	1	1	0	2	0	0	5	2
-31	4	0	1	0	0	1	0	0	0	1	0	0	1
-35	8	1	0	1	0	1	1	2	0	1	0	0	1
-39	7	0	1	1	1	0	0	0	0	0	0	2	2
-40	9	0	0	1	0	2	1	1	0	0	2	2	0
-43	23	1	2	1	3	2	2	2	0	4	1	0	5
Totals		8	18	9	19	19	18	17	14	22	15	17	23

Distribution of primes associated with small discriminants

Table VII

Curve C

D	N_D											
-3	37	3	1	3	3	2	3	1	6	5	4	3	3
-4	16	0	1	0	2	1	1	3	2	2	1	3	0
-7	13	1	0	3	0	1	1	2	2	0	0	1	2
-8	13	1	1	3	1	1	1	1	1	3	0	0	0
-11	15	0	0	2	1	2	3	1	3	1	1	1	0
-15	12	0	1	0	2	1	2	0	2	1	2	0	1
-19	13	0	2	2	2	1	0	0	0	4	1	0	1
-20	8	0	3	0	0	1	2	1	0	0	0	1	0
-23	5	0	0	1	1	2	0	0	0	0	0	0	1
-24	14	0	0	2	1	0	1	4	1	0	1	2	2
-31	7	0	2	0	0	2	0	1	1	1	0	0	0
-35	6	0	1	0	1	0	0	1	1	0	1	0	1
-39	8	0	1	0	2	1	0	0	1	1	0	0	2
-40	4	0	1	0	1	1	0	0	0	0	0	1	0
-43	12	2	1	1	2	1	0	0	0	1	1	3	0
Totals		7	15	17	19	17	14	15	20	19	12	15	13

Curve D

D	N_D											
-3	46	1	2	6	6	3	4	4	4	5	3	4	4
-4	27	3	1	1	1	3	3	0	2	2	4	3	4
-7	14	0	1	1	0	3	0	3	1	3	1	0	1
-8	20	1	1	1	3	4	0	2	2	1	1	3	1
-11	14	0	1	2	1	1	0	2	0	5	2	0	0
-15	11	1	1	1	0	2	1	0	1	2	1	0	1
-19	13	0	0	1	1	2	1	1	2	1	2	0	2
-20	11	0	0	3	1	3	0	1	1	1	0	1	0
-23	5	0	0	0	0	0	2	0	1	1	0	1	0
-24	4	0	0	0	0	1	1	0	0	0	2	0	0
-31	4	0	0	0	1	1	0	0	0	0	0	0	2
-35	9	0	0	0	1	0	1	1	2	1	2	1	0
-39	4	1	1	0	0	0	0	0	1	0	1	0	0
-40	9	1	1	2	1	0	1	1	0	1	0	1	0
-43	6	0	2	1	0	1	0	0	0	0	1	1	0
Totals		8	11	19	16	24	14	15	17	23	20	15	15

Distribution of primes associated with small discriminants

Table VII

Curve $X_o(11)$

D	N_D												
-3	29	1	2	1	3	1	6	4	1	3	2	3	2
-4	67	4	5	4	10	4	7	5	7	7	5	5	4
-7	15	0	1	2	0	3	3	1	0	3	1	1	0
-8	25	0	2	1	0	1	1	2	2	2	3	3	8
-11	88	2	2	7	10	5	10	7	7	13	5	6	14
-15	13	0	0	1	1	2	1	1	2	1	2	0	2
-19	32	1	2	3	1	3	2	1	3	4	4	3	5
-20	11	0	0	1	3	1	0	2	1	1	1	1	0
-23	5	0	0	0	1	0	0	0	1	1	0	0	2
-24	26	1	3	2	2	2	2	1	3	2	4	2	2
-31	19	1	3	1	3	3	0	1	1	2	4	0	0
-35	8	0	0	1	1	1	1	1	0	1	0	0	2
-39	15	0	1	3	1	1	0	2	1	1	1	4	0
-40	5	0	0	1	0	1	0	0	1	0	2	0	0
-43	8	0	0	0	0	1	1	1	1	2	0	1	1
Totals		10	21	28	36	29	34	29	31	43	34	29	42

Distribution of primes associated with small discriminants

EXTENDED RESULTS FOR $X_0(11)$

§5. Discussion and description of tables

As explained below in "remarks on the computations," extended calculations for the curve $X_0(11)$ could be done much more cheaply than for our other examples. We therefore decided to push the calculation further, especially to see how the agreement between conjecture and experiment would come out for this curve and the field $Q(\sqrt{-11})$. The computation was taken as far as

$$p_{189,521} = 2{,}590{,}717$$

for which

$$\pi_{\frac{1}{2}}(2{,}590{,}717) \approx 130.7 \ .$$

The stopping point was quite arbitrary. We wanted to get far enough to have $\pi_{\frac{1}{2}}$ at least four times the value 26.4 reached in the earlier calculation, and it turned out that we could go somewhat further and still stay within our budget.

Table VIII simply lists the supersingular primes and needs no further explanation.

As before, we divided the primes into twelve segments making nearly equal contributions to the sum $\pi_{\frac{1}{2}}(X)$. The division points came out as follows.

Segment	i	p_i
1	615	4523
2	3177	29221
3	8173	83773
4	15860	174299
5	26411	305021
6	39961	479377
7	56616	700949
8	76467	971483
9	99592	1293857
10	126058	1670129
11	155926	2101657
12	189251	2590717

Table IX is just like Table V except that the second group of columns (corresponding to dropping the first two of the twelve segments) now refers to primes p such that

$$p_{3177} = 29221 < p \leq p_{189251} = 2950717 .$$

Table X is like Table VII (referring of course to the segments just listed), except that with the larger frequencies there is some point in including all the discriminants with absolute value less than 100.

Table IX shows good agreement on the whole between the conjectured predictions and actual counts, again with the exception of $Q(\sqrt{-11})$ when the initial segments are not left out. We find it quite reasonable to attribute this discrepancy to slowness of the approach to asymptotic behavior.

There may be identifiable reasons for particularly slow asymptotic approach in this case. Our heuristic arguments involve estimating "probabilities" for coincidences modulo finite M, and then taking a limit as M becomes more and more divisible. The estimates for finite M are arrived at under the assumption that we have primes p with $\sqrt{p} \gg M$. Also we depend on the Tchebotarev density theorem, applied to a Galois extension of degree proportional to M^4, and presumably convergence here is slower when M is large. In many cases the factor we get with a small value of M (say 12) is not very far off the limiting value. For $X_0(11)$, however, the factor 5 makes a big difference and (at least for $D = -11$) the factor 11 is important too. Thus one has to take a fairly large M, and should not expect asymptotic behavior until one reaches very large primes.

It may also be reasonable to expect counts corresponding to very high predicted frequencies to be too low when one looks at an initial segment of primes, simply because there are not enough primes to go around. The relevant factor is the ratio of the number of primes $\leq X$ to $\pi_{\frac{1}{2}}(X)$, and of course it gets larger as X increases. Note that in Table IX the actual frequencies fall noticeably below the predictions whenever the latter are over 100, and that for all the predictions over 150, the probability measure of agreement improves when the initial segments of data are discarded.

We do not offer the above remarks as anything more than suggestive possibilities, but they do reinforce our opinion that the data are consistent with our conjecture.

Table VIII

Curve $X_o(11)$

19	66809	475229	1354649
29	67289	522839	1379659
199	79229	539339	1403239
569	88259	578789	1503149
809	110339	597869	1543319
1289	131479	613999	1562279
1439	150169	628909	1621679
2539	159209	654779	1720109
3319	168869	664109	1728119
3559	196919	666079	1739839
3919	199669	715189	1769539
5519	202109	716299	1792709
9419	209569	724079	1808039
9539	213949	733169	1847999
9929	220279	747449	1875499
11279	226789	749909	1896109
11549	228959	783359	1898759
13229	234869	833509	1981669
14489	243119	860779	2030069
17239	268969	913259	2032799
18149	272999	918839	2077919
18959	273929	920729	2099059
19319	281069	945059	2103389
22279	282889	983429	2122469
24359	289559	1018789	2124359
27529	294989	1024669	2125939
28789	324839	1025939	2146289
32999	325079	1077179	2238529
33029	339539	1088089	2268389
36559	360989	1104659	2405339
42899	372059	1125329	2432719
45259	384089	1147639	2543459
46219	395449	1207439	2547599
49529	402859	1239179	2556739
51169	410009	1269559	2572379
52999	418259	1286099	
55259	464939	1324949	

Supersingular primes ≤ 2,590,717

Table IX

D	N_D	E_D	Pr	N_D	E_D	Pr
-3	165	175.8	.42	141	146.5	.65
-4	376	419.8	.03	318	349.8	.09
-7	86	85.2	.93	72	71.0	.90
-8	88	90.3	.81	74	75.3	.88
-11	528	595.0	.006	461	495.8	.12
-15	63	64.9	.81	52	54.1	.84
-19	185	201.5	.24	162	168.0	.65
-20	56	64.4	.29	46	53.7	.34
-23	40	40.4	.99	37	33.7	.55
-24	134	158.2	.05	113	131.8	.10
-31	100	111.5	.28	81	92.9	.22
-35	54	52.8	.84	48	44.0	.55
-39	88	96.1	.41	77	80.1	.73
-40	48	52.9	.54	43	44.1	.94
-43	44	49.5	.48	37	41.3	.59
-47	27	27.5	.99	25	22.9	.60
-51	110	119.3	.40	85	99.4	.15
-52	41	40.2	.87	37	33.5	.54
-55	32	34.2	.80	29	28.5	.93
-56	96	93.3	.78	87	77.7	.29
-59	91	98.1	.47	77	81.8	.60
-67	39	40.7	.88	33	33.9	.99
-68	27	29.7	.71	24	24.8	.99
-71	61	64.2	.69	53	53.5	.99
-79	64	70.4	.45	57	58.7	.90
-83	28	29.3	.93	26	24.4	.69
-84	86	85.1	.93	74	70.9	.72
-87	28	22.2	.20	27	18.5	.06
-88	11	16.1	.26	11	13.4	.68
-91	99	95.1	.69	86	79.3	.45
-95	23	22.7	.92	20	18.9	.73

Curve $X_o(11)$

Table X

Curve $X_o(11)$

D	N_D											
-3	165	8	16	15	11	16	18	20	15	9	11	16	10
-4	376	26	32	30	24	29	36	36	35	27	32	35	34
-7	86	6	8	1	11	3	6	5	8	8	7	15	8
-8	88	4	10	17	6	8	5	5	9	8	6	3	7
-11	528	26	41	40	39	47	42	47	37	51	62	48	48
-15	63	4	7	6	2	7	7	5	8	4	4	5	4
-19	185	10	13	16	15	17	20	15	19	17	17	12	14
-20	56	5	5	1	7	1	6	5	8	6	1	4	7
-23	40	1	2	3	7	3	2	4	7	5	3	2	1
-24	134	10	11	11	6	12	15	7	9	12	19	15	7
-31	100	11	8	3	10	10	7	14	4	6	8	9	10
-35	54	3	3	6	6	7	3	6	5	5	2	2	6
-39	88	6	5	7	9	9	6	8	8	9	8	5	8
-40	48	2	3	2	8	5	3	7	1	3	3	5	6
-43	44	2	5	5	4	0	3	6	5	2	4	6	2
-47	27	0	2	1	0	2	0	5	4	5	4	3	1
-51	110	10	15	11	7	9	9	4	6	7	11	12	9
-52	41	2	2	3	4	3	3	2	7	1	4	4	6
-55	32	2	1	0	1	7	4	3	1	1	8	3	1
-56	96	5	4	10	5	12	8	8	4	10	9	12	9
-59	91	6	8	8	13	9	7	6	3	7	7	7	10
-67	39	4	2	4	1	4	4	0	3	6	2	4	5
-68	27	1	2	0	3	0	3	4	2	1	1	5	5
-71	61	3	5	7	7	5	5	8	6	4	1	4	6
-79	64	2	5	8	3	3	7	5	10	6	6	2	7
-83	28	1	1	2	5	1	2	5	2	1	1	4	3
-84	86	5	7	9	6	8	8	5	5	9	10	5	9
-87	28	1	0	2	2	2	2	4	5	2	3	4	1
-88	11	0	0	0	0	1	0	1	1	1	3	0	4
-91	99	3	10	13	10	7	7	5	7	6	12	8	11
-95	23	0	3	2	3	3	1	2	6	0	0	1	2
Totals		169	236	243	235	250	249	257	250	239	269	260	261

Distribution of primes associated with discriminants ≥ -95.

Remarks on the computations

We conclude with some comments on the machine computations. For $X_0(11)$, t_p was found as the coefficient of q^p in the q-expansion

$$q \prod_{n=1}^{\infty} (1-q^n)^2 (1-q^{11n})^2 .$$

Cf. Shimura [Sh] and Ligozat [Li], where such products are given also for other modular curves. Euler's formula gives

$$\prod_{n=1}^{\infty} (1-q^n)$$

as a series in which all coefficients are zero except for 1's and −1's occurring for n in certain quadratic progressions. Straightforward multiplication of the series is an easy computation if one wants only 50,000 or so terms.

For the other curves, the calculation was based on the formula

$$t_p = 1 + p - N_p ,$$

where N_p is the number of rational points on the curve over F_p. The equation of the curve can be rewritten if necessary to have the form

$$y^2 = f(x) = 4x^3 + a_2 x^2 + a_4 x + a_6 ,$$

without changing the value of N_p for any $p > 2$. The calculation is then done by first constructing a table of values

$$r(0), r(1), \cdots, r(p-1) ,$$

where

$$r(i) = 1 + \left(\frac{i}{p}\right) ,$$

i.e. the number of solutions of $y^2 = r(i) \bmod p$ [Sw − D 2]. Then

$$N_p - 1 = \sum_{i=1}^{p} r(f(i)).$$

It is worth noting that successive values of $f(i)$ (and the quadratic residues needed to construct the table $r(i)$), can be built up by successive addition from constant third (second) differences. By checking each number occurring in the calculation when it is formed, reduction mod p can always be carried out by a single addition or subtraction of p. This avoids the comparatively slow machine operation of integer division, and leads to significantly faster calculation than the more obvious method of calculating $f(x)$ by multiplications and additions, followed by reduction mod p. The computing time needed to calculate t_p by this method is clearly proportional to p, so the cost of finding t_p for all $p \leq x$ rises rapidly with x. We picked the 5,000th prime as a stopping point as a reasonable compromise between cost of computation and the need for a significant sample of data.

The equation for the points of order 5 on $X_0(11)$ was found using a program which calculates the polynomials for the n-th division points (x-coordinate), given in most books, e.g. in Weber, and sometimes referred to as $\psi_n(x)$. It was factored with the help of a program which uses a version of Berlekamp's algorithm for factoring polynomials modulo primes, cf. [B].

In the more extended calculations for $X_0(11)$, the relevant terms of the series could not all be kept in core storage at one time, so a new program was required to use auxiliary storage for intermediate results. By far the greatest part of the computer time was spent on the final multiplication

$$\prod (1-q^n) \times \prod (1-q^n)(1-q^{11n})^2.$$

As noted above, the series for $\prod (1-q^n)$ has very few non-zero terms (which is what makes the calculation feasible at all). In the factor on the right, the density of non-zero coefficients turns out to be a little over one-half and (perhaps surprisingly) to be increasing steadily, though very slowly, as one goes to higher order terms.

BIBLIOGRAPHY

[B] E. R. BERLEKAMP, Algebraic coding theory, McGraw Hill, N.Y., 1968.

[H-L] G. H. HARDY and J. LITTLEWOOD, Partitio Numerorum, Acta Math. 44 (1923), pp. 1-70, especially p. 48.

[H-C] HARISH-CHANDRA, Harmonic analysis on reductive p-adic groups, Proceedings of symposia in pure mathematics, AMS, Williamstown conference (1973), pp. 167-192.

[I] Y. IHARA, Hecke polynomials as congruence zeta functions in elliptic modular case, Ann. of Math. 85 (1967), pp. 267-295.

[L 1] S. LANG, Elliptic functions, Addison Wesley, Reading, 1973.

[L 2] ―――, Algebraic Number Theory, Addison Wesley, Reading, 1971

[Le] D. H. LEHMER, Note on the distribution of Ramanujan's Tau function, Math. Comp. 24 (1970), pp. 741-743.

[Li] G. LIGOZAT, Courbes modulaires de genre 1, to appear.

[Ma] B. MAZUR, Rational points of abelian varieties with values in towers of number fields, Inv. Math. 18 (1972), pp. 183-266.

[Man] J. MANIN, Periods of parabolic forms and p-adic Hecke series, Mat. Sbornik n.s. 92 (1973), 378-401, English translation, Math. USSR Sbornik 21 (1973), 371-393.

[R] K. RIBET, On ℓ-adic representations attached to modular forms, to appear.

[Sa-Sh 1] SALLY and J. SHALIKA, Characters of the discrete series of representations of SL(2) over a local field, Proc. Nat. Acad. Sci. USA 61 (1968), pp. 1231-1237.

[Sa-Sh 2] ―――, The Plancherel formula for SL(2) over a local field, Proc. Nat. Acad. Sci. USA 63 (1969), pp. 661-667.

[Sa-Sh 3] ―――, The Fourier transform on SL_2 over a non-archimedean local field, to appear.

[S 1] J. P. SERRE, Groupes de Lie ℓ-adiques attachés aux courbes élliptiques, Colloque Clermont-Ferrand, Les tendances géomètriques en algèbre et théorie des nombres, 1964, sections 3.4 and 4.3.

[S 2]   J. P. SERRE, Propriétés Galoisiennes des points d'ordre fini des courbes élliptiques, Inv. Math. 15 (1972), pp. 259-331.

[S 3]   _____._____, Abelian ℓ-adic representations and elliptic curves, Benjamin, 1968.

[Sh]   G. SHIMURA, A reciprocity law in non-solvable extensions, J. reine angew. Math. 221 (1966), pp. 209-220.

[SwD]   P. SWINNERTON-DYER, On ℓ-adic representations and congruences for coefficients of modular forms, Springer Lecture Notes 350 (Antwerp conference).

[SwD 2]   _____, An application of computing to class field theory, in Algebraic Number Theory (Brighton Conference, edited by J. W. S. Cassels and A. Frölich) Thompson Book Co., Washington, 1967.

[T]   J. TATE, The Arithmetic of elliptic curves, Invent. Math. 23 (1974), pp. 179-206.

[Tu]   T. A. TUSKINA, A numerical experiment on the calculation of the Hasse invariant for certain curves, Izv. Akad. Nauk SSSR Ser. Mat. 29 (1965), 1203-1204, English translation, AMS Translations (Ser. 2) 66, (1968), 204-205.

[Y]   H. YOSHIDA, On an analogue of the Sato conjecture, Invent. Math. 19 (1973), pp. 261-277.

Vol. 342: Algebraic K-Theory II, "Classical" Algebraic K-Theory, and Connections with Arithmetic. Edited by H. Bass. XV, 527 pages. 1973.

Vol. 343: Algebraic K-Theory III, Hermitian K-Theory and Geometric Applications. Edited by H. Bass. XV, 572 pages. 1973.

Vol. 344: A. S. Troelstra (Editor), Metamathematical Investigation of Intuitionistic Arithmetic and Analysis. XVII, 485 pages. 1973.

Vol. 345: Proceedings of a Conference on Operator Theory. Edited by P. A. Fillmore. VI, 228 pages. 1973.

Vol. 346: Fučík et al., Spectral Analysis of Nonlinear Operators. II, 287 pages 1973.

Vol. 347: J. M. Boardman and R. M. Vogt, Homotopy Invariant Algebraic Structures on Topological Spaces. X, 257 pages. 1973.

Vol. 348: A. M. Mathai and R. K. Saxena, Generalized Hypergeometric Functions with Applications in Statistics and Physical Sciences. VII, 314 pages. 1973.

Vol. 349: Modular Functions of One Variable II Edited by W. Kuyk and P. Deligne V, 598 pages. 1973

Vol. 350: Modular Functions of One Variable III. Edited by W. Kuyk and J.-P. Serre. V, 350 pages. 1973.

Vol 351: H Tachikawa, Quasi-Frobenius Rings and Generalizations XI, 172 pages. 1973

Vol. 352: J. D. Fay, Theta Functions on Riemann Surfaces. V, 137 pages 1973

Vol. 353: Proceedings of the Conference on Orders, Group Rings and Related Topics. Organized by J. S. Hsia, M. L. Madan and T. G. Ralley. X, 224 pages. 1973.

Vol. 354: K. J Devlin, Aspects of Constructibility. XII, 240 pages. 1973.

Vol. 355: M Sion, A Theory of Semigroup Valued Measures. V, 140 pages. 1973.

Vol. 356: W. L. J. van der Kallen, Infinitesimally Central Extensions of Chevalley Groups VII, 147 pages 1973.

Vol. 357: W. Borho, P Gabriel und R. Rentschler, Primideale in Einhüllenden auflösbarer Lie-Algebren. V, 182 Seiten. 1973.

Vol. 358: F. L. Williams, Tensor Products of Principal Series Representations. VI, 132 pages. 1973.

Vol 359: U Stammbach, Homology in Group Theory. VIII, 183 pages. 1973

Vol. 360: W. J. Padgett and R. L. Taylor, Laws of Large Numbers for Normed Linear Spaces and Certain Fréchet Spaces. VI, 111 pages. 1973

Vol. 361: J. W. Schutz, Foundations of Special Relativity: Kinematic Axioms for Minkowski Space-Time. XX, 314 pages. 1973.

Vol 362: Proceedings of the Conference on Numerical Solution of Ordinary Differential Equations. Edited by D.G. Bettis. VIII, 490 pages. 1974.

Vol. 363: Conference on the Numerical Solution of Differential Equations. Edited by G. A. Watson. IX, 221 pages. 1974.

Vol. 364: Proceedings on Infinite Dimensional Holomorphy. Edited by T. L. Hayden and T. J. Suffridge. VII, 212 pages. 1974.

Vol. 365: R. P. Gilbert, Constructive Methods for Elliptic Equations. VII, 397 pages. 1974.

Vol. 366: R. Steinberg, Conjugacy Classes in Algebraic Groups (Notes by V. V. Deodhar). VI, 159 pages. 1974.

Vol. 367: K. Langmann und W. Lütkebohmert, Cousinverteilungen und Fortsetzungssätze. VI, 151 Seiten. 1974.

Vol. 368: R. J. Milgram, Unstable Homotopy from the Stable Point of View. V, 109 pages. 1974.

Vol. 369: Victoria Symposium on Nonstandard Analysis. Edited by A. Hurd and P. Loeb. XVIII, 339 pages. 1974.

Vol. 370: B. Mazur and W. Messing, Universal Extensions and One Dimensional Crystalline Cohomology. VII, 134 pages. 1974.

Vol. 371: V Poenaru, Analyse Différentielle V, 228 pages. 1974

Vol. 372: Proceedings of the Second International Conference on the Theory of Groups 1973. Edited by M. F. Newman. VII, 740 pages. 1974.

Vol 373: A E R Woodcock and T. Poston, A Geometrical Study of the Elementary Catastrophes. V, 257 pages. 1974.

Vol 374: S Yamamuro, Differential Calculus in Topological Linear Spaces IV, 179 pages 1974.

Vol. 375: Topology Conference. Edited by R. F. Dickman Jr. and P Fletcher. X, 283 pages 1974.

Vol. 376: I. J. Good and D. B. Osteyee, Information, Weight of Evidence. The Singularity between Probability Measures and Signal Detection XI, 156 pages. 1974.

Vol 377: A M Fink, Almost Periodic Differential Equations. VIII, 336 pages 1974

Vol. 378: TOPO 72 – General Topology and its Applications Proceedings 1972. Edited by R. A. Alò, R. W. Heath and J. Nagata XIV, 651 pages. 1974.

Vol. 379: A. Badrikian et S. Chevet, Mesures Cylindriques, Espaces de Wiener et Fonctions Aléatoires Gaussiennes X, 383 pages. 1974.

Vol 380: M. Petrich, Rings and Semigroups. VIII, 182 pages. 1974.

Vol. 381: Séminaire de Probabilités VIII. Edité par P. A. Meyer. IX, 354 pages. 1974.

Vol. 382: J. H. van Lint, Combinatorial Theory Seminar Eindhoven University of Technology. VI, 131 pages 1974

Vol. 383: Séminaire Bourbaki - vol 1972/73. Exposés 418-435 IV, 334 pages 1974

Vol. 384: Functional Analysis and Applications, Proceedings 1972. Edited by L. Nachbin. V, 270 pages. 1974.

Vol. 385: J. Douglas Jr. and T. Dupont, Collocation Methods for Parabolic Equations in a Single Space Variable (Based on C^1-Piecewise-Polynomial Spaces). V, 147 pages. 1974.

Vol 386: J Tits, Buildings of Spherical Type and Finite BN-Pairs. X, 299 pages 1974.

Vol. 387: C. P. Bruter, Eléments de la Théorie des Matroïdes V, 138 pages 1974.

Vol. 388: R. L. Lipsman, Group Representations. X, 166 pages. 1974

Vol. 389: M.-A. Knus et M. Ojanguren, Théorie de la Descente et Algèbres d' Azumaya. IV, 163 pages. 1974.

Vol. 390: P A Meyer, P. Priouret et F. Spitzer, Ecole d'Eté de Probabilités de Saint-Flour III – 1973. Edité par A. Badrikian et P.-L. Hennequin. VIII, 189 pages. 1974

Vol. 391: J W. Gray, Formal Category Theory: Adjointness for 2-Categories. XII, 282 pages. 1974.

Vol. 392: Géométrie Différentielle, Colloque, Santiago de Compostela, Espagne 1972. Edité par E. Vidal. VI, 225 pages. 1974

Vol. 393: G. Wassermann, Stability of Unfoldings. IX, 164 pages 1974.

Vol. 394: W. M. Patterson, 3rd, Iterative Methods for the Solution of a Linear Operator Equation in Hilbert Space – A Survey. III, 183 pages. 1974.

Vol. 395: Numerische Behandlung nichtlinearer Integrodifferential- und Differentialgleichungen. Tagung 1973 Herausgegeben von R. Ansorge und W. Törnig. VII, 313 Seiten. 1974

Vol. 396: K. H. Hofmann, M. Mislove and A. Stralka, The Pontryagin Duality of Compact O-Dimensional Semilattices and its Applications. XVI, 122 pages 1974

Vol 397: T. Yamada, The Schur Subgroup of the Brauer Group V, 159 pages. 1974.

Vol 398: Théories de l'Information, Actes des Rencontres de Marseille-Luminy, 1973 Edité par J. Kampé de Fériet et C -F Picard. XII, 201 pages. 1974.

Vol. 399: Functional Analysis and its Applications. Proceedings 1973. Edited by H. G. Garnir, K. R. Unni and J. H. Williamson. II, 584 pages. 1974.

Vol. 400: A Crash Course on Kleinian Groups. Proceedings 1974. Edited by L. Bers and I. Kra. VII, 130 pages. 1974.

Vol. 401: M. F. Atiyah, Elliptic Operators and Compact Groups. V, 93 pages. 1974.

Vol. 402: M. Waldschmidt, Nombres Transcendants. VIII, 277 pages. 1974.

Vol. 403: Combinatorial Mathematics. Proceedings 1972. Edited by D. A. Holton. VIII, 148 pages. 1974.

Vol. 404: Théorie du Potentiel et Analyse Harmonique. Edité par J. Faraut. V, 245 pages. 1974.

Vol. 405: K. J. Devlin and H. Johnsbråten, The Souslin Problem. VIII, 132 pages. 1974.

Vol. 406: Graphs and Combinatorics. Proceedings 1973. Edited by R. A. Bari and F. Harary. VIII, 355 pages. 1974.

Vol. 407: P. Berthelot, Cohomologie Cristalline des Schémas de Caracteristique p > o. II, 604 pages. 1974.

Vol. 408: J. Wermer, Potential Theory. VIII, 146 pages. 1974.

Vol. 409: Fonctions de Plusieurs Variables Complexes, Séminaire François Norguet 1970-1973. XIII, 612 pages. 1974.

Vol. 410: Séminaire Pierre Lelong (Analyse) Année 1972-1973. VI, 181 pages. 1974.

Vol. 411: Hypergraph Seminar. Ohio State University, 1972. Edited by C. Berge and D. Ray-Chaudhuri. IX, 287 pages. 1974.

Vol. 412: Classification of Algebraic Varieties and Compact Complex Manifolds. Proceedings 1974. Edited by H. Popp. V, 333 pages. 1974.

Vol. 413: M. Bruneau, Variation Totale d'une Fonction. XIV, 332 pages. 1974.

Vol. 414: T. Kambayashi, M. Miyanishi and M. Takeuchi, Unipotent Algebraic Groups. VI, 165 pages. 1974.

Vol. 415: Ordinary and Partial Differential Equations. Proceedings 1974. XVII, 447 pages. 1974.

Vol. 416: M. E. Taylor, Pseudo Differential Operators. IV, 155 pages. 1974.

Vol. 417: H. H. Keller, Differential Calculus in Locally Convex Spaces. XVI, 131 pages. 1974.

Vol. 418: Localization in Group Theory and Homotopy Theory and Related Topics. Battelle Seattle 1974 Seminar. Edited by P. J. Hilton. VI, 172 pages 1974.

Vol. 419: Topics in Analysis. Proceedings 1970. Edited by O. E. Lehto, I. S. Louhivaara, and R. H. Nevanlinna. XIII, 392 pages. 1974.

Vol. 420: Category Seminar. Proceedings 1972/73. Edited by G. M. Kelly. VI, 375 pages. 1974.

Vol. 421: V. Poénaru, Groupes Discrets. VI, 216 pages. 1974.

Vol. 422: J.-M. Lemaire, Algèbres Connexes et Homologie des Espaces de Lacets. XIV, 133 pages. 1974.

Vol. 423: S. S. Abhyankar and A. M. Sathaye, Geometric Theory of Algebraic Space Curves. XIV, 302 pages. 1974.

Vol. 424: L. Weiss and J. Wolfowitz, Maximum Probability Estimators and Related Topics. V, 106 pages. 1974.

Vol. 425: P. R. Chernoff and J. E. Marsden, Properties of Infinite Dimensional Hamiltonian Systems. IV, 160 pages. 1974.

Vol. 426: M. L. Silverstein, Symmetric Markov Processes. X, 287 pages. 1974.

Vol. 427: H. Omori, Infinite Dimensional Lie Transformation Groups. XII, 149 pages. 1974.

Vol. 428: Algebraic and Geometrical Methods in Topology, Proceedings 1973. Edited by L. F. McAuley. XI, 280 pages. 1974.

Vol. 429: L. Cohn, Analytic Theory of the Harish-Chandra C-Function. III, 154 pages. 1974.

Vol. 430: Constructive and Computational Methods for Differential and Integral Equations. Proceedings 1974. Edited by D. L. Colton and R. P. Gilbert. VII, 476 pages. 1974.

Vol. 431: Séminaire Bourbaki - vol. 1973/74. Exposés 436-452. IV, 347 pages. 1975.

Vol. 432: R. P. Pflug, Holomorphiegebiete, pseudokonvexe Gebiete und das Levi-Problem. VI, 210 Seiten. 1975.

Vol. 433: W. G. Faris, Self-Adjoint Operators. VII, 115 pages. 1975.

Vol. 434: P. Brenner, V. Thomée, and L. B. Wahlbin, Besov Spaces and Applications to Difference Methods for Initial Value Problems. II, 154 pages. 1975.

Vol. 435: C. F. Dunkl and D. E. Ramirez, Representations of Commutative Semitopological Semigroups. VI, 181 pages. 1975.

Vol. 436: L. Auslander and R. Tolimieri, Abelian Harmonic Analysis, Theta Functions and Function Algebras on a Nilmanifold. V, 99 pages. 1975.

Vol. 437: D. W. Masser, Elliptic Functions and Transcendence. XIV, 143 pages. 1975.

Vol. 438: Geometric Topology. Proceedings 1974. Edited by L. C. Glaser and T. B. Rushing. X, 459 pages. 1975.

Vol. 439: K. Ueno, Classification Theory of Algebraic Varieties and Compact Complex Spaces. XIX, 278 pages. 1975

Vol. 440: R. K. Getoor, Markov Processes: Ray Processes and Right Processes. V, 118 pages. 1975.

Vol. 441: N. Jacobson, PI-Algebras. An Introduction. V, 115 pages. 1975.

Vol. 442: C. H. Wilcox, Scattering Theory for the d'Alembert Equation in Exterior Domains. III, 184 pages. 1975.

Vol. 443: M. Lazard, Commutative Formal Groups. II, 236 pages. 1975.

Vol. 444: F. van Oystaeyen, Prime Spectra in Non-Commutative Algebra. V, 128 pages. 1975.

Vol. 445: Model Theory and Topoi. Edited by F. W. Lawvere, C. Maurer, and G. C. Wraith. III, 354 pages. 1975.

Vol. 446: Partial Differential Equations and Related Topics. Proceedings 1974. Edited by J. A. Goldstein. IV, 389 pages. 1975.

Vol. 447: S. Toledo, Tableau Systems for First Order Number Theory and Certain Higher Order Theories. III, 339 pages. 1975.

Vol. 448: Spectral Theory and Differential Equations. Proceedings 1974. Edited by W. N. Everitt. XII, 321 pages. 1975.

Vol. 449: Hyperfunctions and Theoretical Physics. Proceedings 1973. Edited by F. Pham. IV, 218 pages. 1975.

Vol. 450: Algebra and Logic. Proceedings 1974. Edited by J. N. Crossley. VIII, 307 pages. 1975.

Vol. 451: Probabilistic Methods in Differential Equations. Proceedings 1974. Edited by M. A. Pinsky. VII, 162 pages. 1975.

Vol. 452: Combinatorial Mathematics III. Proceedings 1974. Edited by Anne Penfold Street and W. D. Wallis. IX, 233 pages. 1975.

Vol. 453: Logic Colloquium. Symposium on Logic Held at Boston, 1972-73. Edited by R. Parikh. IV, 251 pages. 1975.

Vol. 454: J. Hirschfeld and W. H. Wheeler, Forcing, Arithmetic, Division Rings. VII, 266 pages. 1975.

Vol. 455: H. Kraft, Kommutative algebraische Gruppen und Ringe. III, 163 Seiten. 1975.

Vol. 456: R. M. Fossum, P. A. Griffith, and I. Reiten, Trivial Extensions of Abelian Categories. Homological Algebra of Trivial Extensions of Abelian Categories with Applications to Ring Theory. XI, 122 pages. 1975.

MIX
Papier aus verantwortungsvollen Quellen
Paper from responsible sources
FSC® C105338

If you have any concerns about our products,
you can contact us on
ProductSafety@springernature.com

In case Publisher is established outside the EU,
the EU authorized representative is:
**Springer Nature Customer Service Center GmbH
Europaplatz 3, 69115 Heidelberg, Germany**

Printed by Libri Plureos GmbH
in Hamburg, Germany